Security, Conflict and Cooperation in the Contemporary World

The Palgrave Macmillan series, Security, Conflict and Cooperation in the Contemporary World aims to make a significant contribution to academic and policy debates on international history, war and diplomacy since 1900. It evolved from the series Global Conflict and Security edited by Professor Saki Ruth Dockrill. The current series welcomes proposals that offer innovative historical perspectives, based on archival evidence and promoting an empirical understanding of economic and political cooperation, conflict and security, peace-making, diplomacy, humanitarian intervention, nation-building, intelligence, terrorism, the influence of ideology and religion on international relations, as well as the work of international organisations and non-governmental organisations.

Gareth Michael Jones

The Development of Nuclear Propulsion in the Royal Navy, 1946–1975

palgrave
macmillan

Gareth Michael Jones
Ugborough, Devon, UK

ISSN 2731-6807 ISSN 2731-6815 (electronic)
Security, Conflict and Cooperation in the Contemporary World
ISBN 978-3-031-05128-9 ISBN 978-3-031-05129-6 (eBook)
https://doi.org/10.1007/978-3-031-05129-6

Cover credit: Keystone Press / Alamy Stock Photo

This Palgrave Macmillan imprint is published by the registered company Springer Nature Switzerland AG.
The registered company address is: Gewerbestrasse 11, 6330 Cham, Switzerland

Preface

The Royal Navy's first submarine, *Holland 1*, was launched in 1901. In that same year H. G. Wells wrote: '…my imagination, in spite even of spurring, refuses to see any sort of submarine doing anything but suffocate its crew and floundering at sea'.[1] *Holland 1* was powered on the surface by a petrol engine, and was a submarine with very little freeboard. When dived, she was propelled by lead acid batteries driving an electric motor. With no air purification systems to support the crew of eight men, white mice were carried onboard to warn against poisonous fumes. It is not difficult, therefore, to understand Wells' imagination visualising anything other than disaster. However, his contemporary author, Jules Verne, saw things differently, especially in Captain Nemo's *Nautilus.* Whilst taking Professor Aronnax on a tour of the *Nautilus*, Nemo tells his captive that the submarine is powered by "Bunsen's contrivances" and that: '…the dynamic power of my engines is almost infinite'.[2] *Holland 1* was a submersible that transited on the surface and dived to attack its target. Verne's vision in that of the *Nautilus* is of a true submarine diving to great depths and travelling at great speed. Verne's vision of 1873 eventually became a reality eighty-two years later in 1955, when the nuclear-powered submarine, USS *Nautilus*, sailed on her maiden sea trials. This is a book which

[1] H G Wells, *Anticipations of the Reaction of Mechanical and Scientific Progress upon Human Life and Thought* (Chapman and Hall, London, 1901) Quoted in, https://en.wikipedia.org/wiki/Anticipations

[2] Jules Verne, *Twenty Thousand Leagues Under the Sea* (Butler Brothers, New York and Chicago, 1887) p. 78. https://www.google.co.uk/books/edition/Twenty_Thousand_Leagues_Under_the_Sea

investigates the Royal Navy's journey in the third quarter of the last century to achieve the true submarine, able to sail the ocean depths free from the constraints of the atmosphere, through the Naval Nuclear Propulsion Programme; the development of nuclear propulsion for its submarine fleet; and thereby its ability to effectively counter the Soviet submarine threat of the Cold War and carry the nation's nuclear deterrent.

Ugborough, UK　　Gareth Michael Jones

It is an immense privilege, both to have been so closely involved with the UK's nuclear submarine programme and to have been invited to pen an introduction to this fascinating record of the development of nuclear propulsion through to 1975—the year before I joined the Royal Navy.

Having spent almost the entirety of my career either in or closely associated with the Submarine Service, although I had a general understanding of the circumstances leading to the advent of the UK's first nuclear submarine, HMS Dreadnought, I had little appreciation of the details underpinning the ebb and flow of events that defined this journey.

'Taff' Jones' research has helped to bring to life a timely narrative, recording the importance of vision, leadership, the demand for technical excellence, the relationship between the US and the UK and the all important component of 'people'—their characters, their motivations and the historical environment in which those involved played out their roles. While exploring the difficulties encountered on the journey through the early years of the programme, it is ultimately an account of success, the foundation on which the effectiveness of today's Royal Naval submarines has been built.

As the search for further technological improvements continues, the lessons learnt along the way remain as relevant today as when first encountered—a feature recognised by Churchill in stating that 'the further you look back the more you can see into the future'. While the nuclear submarine represents a highly potent war-fighting capability, it must continue to enshrine a balance of safety, quality and reliability executed through a cadre of competent designers, manufacturers, maintainers and operators who, together, are entrusted with the effective stewardship of this technology.

Despite the hugely attractive advantage that this nuclear-powered underwater capability represents, it is no coincidence that there are so few nations that have proved to have the necessary willpower, technological base and competences necessary to establish and sustain such a complex programme. For the UK, this represents both national and international endeavours that demand enduring clarity of the goal and agreement of its 'worth'; aligned prioritisation of effort; appropriate funding; honesty and realism; and, above all, competent people working in a structured organisation that enjoys inspired, effective and accountable leadership.

This review of the evolution of the UK's nuclear submarine programme is both a timely reminder of the early steps that were undertaken on the journey and an illustration of the focused effort it will take to deliver continued success by those to whom the baton of effective stewardship has been passed. Rear Admiral Steve Lloyd CBE (Retired Marine Engineer Submarines).

Acknowledgements

To write a nuclear history is a difficult undertaking due to the secrecy surrounding the subject matter. It would have been more difficult without the introduction to many people involved in the nascent days of the Royal Navy's Naval Nuclear Propulsion Programme who gave generously of their time and, on occasions, hospitality. First and foremost, I offer my sincere gratitude to Rear Admiral Steve Lloyd CBE, who listened to my initial thoughts and supported my research from the start. Rear Admiral Lloyd subsequently introduced me to Vice Admiral Sir Robert Hill who in turn introduced me to Rear Admiral Peter (Spam) Hammersley, Captain Colin Farley-Sutton RN Ret'd, Captain John Jacobsen RN Ret'd and Commander Roger Berry RN Ret'd; I was later introduced to former CPO W. (Baz) Bowyer. I thank them all for answering my numerous correspondences, for their advice and their unfailing support for my research.

My sincere gratitude goes to Commodore Mark Adams RN, Director Nuclear Propulsion (DNP), for his support and granting me privileged access to unreleased files held at his offices at MoD Abbey Wood. Members of DNP's staff meriting an expression of gratitude include DNP's former Secretary, Mrs Sandy Grinnall for organising my visits and finding me a desk to work from, Ms Aimee Pugh for dealing with subsequent queries and Mr Paul Bolt, DNP's Security Information Policy Manager, for his advice. My supervisor and Director of Studies, Dr Harry Bennett has my deep gratitude for his light touch, wide knowledge of naval history and gratefully received advice and support. I also offer my deep gratitude to Dr Elaine Murphy for her encouragement, support and advice.

Final thanks go to all my friends for their support but especially my “student widow” wife, Philomena, for her love, encouragement and understanding over the past six years.

Dramatis Personae

Ackers, Wallace A.,	Director, Tube Alloys
Alexander, A. V.	First Lord of the Admiralty, 1940–46; Minister of Defence, 1946–50
Anderson, Sir John,	Chancellor of the Exchequer, 1943–45
Baker, Rowland, RCNC	Technical Chief Executive DPT
Barman, H. L.,	Rolls-Royce Ltd, Senior Chief Executive
Berry, Roger, Commander	Harwell & DLE, Groton USA
Buckley, J. W., Major	Metropolitan-Vickers Ltd
Burke, Arleigh, Admiral USN	Chief of Naval Operations, 1955–61
Brundrett, Sir Frederick,	Chairman DRPC
Caccia, Sir Harold	British Ambassador to Washington
Carroll, Dr John A.,	Deputy Controller, 1946–58
Cilcennan, Viscount	First Lord of the Admiralty, 1951–56
Cockcroft, Sir John,	Harwell, AERE Director
Cotman, D. A., Captain	DLE, Pittsburgh USA
Daniel, Sir Charles, Vice Admiral	Controller of the Navy, 1945–49
Daniel, R. J., RCNC	Chief Constructor DPT, 1960
Denny, Sir Michael, Admiral	Controller of the Navy, 1949–53
Diamond, Jack, RNSS	Head of Naval Section Harwell, 1946–53
Dunphie, Charles, Major General	Vickers-Armstrongs, Chairman
Dunworth, J. V.,	Harwell, Head Nuclear Physics Division
Edwards, Prof. J., RNSS	Harwell then Greenwich, 1957–
Edwards, Sir Ralph, Admiral	Controller of the Navy, 1953–56

Eisenhower, Dwight D.,	President USA, 1953–61
Elkins, R. F., Vice Admiral	(BJSM) Washington, 1957–58
Farley-Sutton, Colin,	Captain, Dreadnought Project Team
Finniston, H. M.,	Harwell, Head of Metallurgy
Goodlet, B. L.,	Harwell, Engineering R&D Division
Gordon, Robert,	Westinghouse
Hall, Viscount	First Lord of the Admiralty, 1946–51
Hammersley, Peter, Rear Admiral	HMS *Dreadnought*
Harrison-Smith, S., Captain	Head of Naval Section, 1954–58
Hawkins, R. S., Rear Admiral	Director General Ships
Head, Anthony	Minister of Defence, 1956–57
Heathcoat-Amory, Derick	Chancellor of the Exchequer, 1958–60
Hedger, H. E.,	Admiralty, Asst. Director of Contracts
Heslop, J. A. B.,	Admiralty, Seconded to AERE
Hinton, Sir Christopher,	Risley, Industrial Group Director
Hives, Lord	Rolls-Royce Ltd, Chairman
Hudspith, H. C.,	UKAEA, Finance
Jacobsen, John, Captain	HMS *Valiant*
Kingcombe, Sir John,	Vice Admiral, Engineer-in-Chief, 1945–47
Lang, Sir John,	Secretary of the Admiralty, 1947–61
Le Fanu, Sir Michael,	Admiral, Controller of the Navy, 1961–65
Mackay, J. M.,	Admiralty, Under Secretary Finance
MacLean, I. G., Rear Admiral	Deputy E-in-C 1953–55
Macmillan, Harold	Prime Minister, 1957–63
Mason, Sir Frank T., Vice Admiral	Engineer-in-Chief, 1953–56
Mawer, F. H.,	Admiralty, General Finance Branch
McKinnell,	Admiralty, General Finance Branch
McMahon, Brien, US Senator	Sponsor of US Atomic Energy Act, 1946
Mountbatten, Earl, Admiral of the Fleet	First Sea Lord, 1951–55
Orem, H. J.,	HM Treasury
Owen, W. L.,	Risley, Industrial Group Deputy Director
Peirson, D. E. H.,	UKAEA, Secretary
Penney, Sir William,	UKAEA, Aldermaston Director
Plowden, Lord	UKAEA Chairman
Portal, Lord	(MoS) Controller of Atomic Energy
Powell, Sir Richard P.,	(MoD) Permanent Secretary
Price, B. Terence,	Harwell, AERE Physicist

Pritchard, A. A.,	Admiralty, Material Division
Radice, I. D.,	HM Treasury
Rebbeck, Sir Edward, Rear Admiral	Vickers-Armstrongs Group
Reid, Sir Peter, Admiral	Controller of the Navy, 1956–61
Rickover, Hyman G., Admiral USN	Nuclear Programme 1946–82
Ridley, W. T. C., Captain	Dreadnought Project Team
Rydill, L. J., RCNC	HMS *Dreadnought* designer
Samuel, H. W.,	Admiralty, Under Secretary Finance
Sandys, Duncan,	Minister of Defence, 1957–59
Sarell, R. I. A., Captain	E-in-C's Department
Schonland, Dr Basil F. J.,	Harwell, AERE Deputy Director
Selkirk, Earl of	First Lord of the Admiralty, 1957–59
Serpell, D. R.,	HM Treasury
Sims, Sir Alfred J., RCNC	Director General Ships, 1958–67
Stewart, J. C. C.,	Risley, Industrial Group
Strath, Sir William,	UKAEA, Managing Director, 1955–59
Synott, P. N. N.,	Admiralty, Under Secretary Finance
Thorneycroft, Peter,	Chancellor of the Exchequer, 1957–58
Tizard, Sir Henry,	Chairman DRPC, 1950–
Tongue, H.,	Harwell, AERE Chief Engineer
Verity, C. E. H.,	Foster Wheeler Ltd
Weeks, Sir Ronald,	Vickers-Armstrongs Group, Chairman
White, E. A.,	Risley, Industrial Group (H&S Branch)
Willis, J. A. V.,	UKAEA representative, Washington DC
Wilson, G. A. M., Rear Admiral	(RANP), 1957–59
Yapp, R. P. H.,	Vickers-Armstrongs
Zuckerman, Sir Solly,	Government Scientific Advisor

Chronology

1939, 17 Dec. (Germany) Discovery of fission of uranium by Otto Hahn

1940, Mar. (UK) Frisch-Peierls Memorandum

1941, 15 Jul. (UK) MAUD Committee Report on nuclear fission released

1942, 2 Dec. (US) First controlled nuclear chain reaction in the Chicago Pile

1943, (UK) Secondment of Prof. Mark Oliphant and other RNSS staff to the Tube Alloys Project

1945, Nov. (UK) Atomic Energy Research Establishment, Harwell founded

1946, 1 Jan. (UK) Secondment of Jack Diamond as head of the Naval Section at Harwell

1946, 18 Feb. (UK) Rear Admiral Charles Daniel's paper on nuclear propulsion

1946, 10 Apr. (US) Paper by A. M. Weinberg in the US identifying the Pressurised Water Reactor as best suited for use in a submarine

1946, Jun. (US) Captain H. G. Rickover USN appointed to Oak Ridge National Laboratory to head USN team to investigate nuclear propulsion

1948, 14 Nov. (UK) Formation of Enriched Reactor Group for submarine prototype

1950, 6 Sep. (UK) Contract awarded to Metropolitan-Vickers to confirm the feasibility of a nuclear submarine based on the Mark I Enriched Reactor

1951, Jul. (US) USS *Nautilus* authorised by the US Congress

1952, May (UK) Conclusion that the Enriched Reactor would be too large for a submarine

1953, Jun. (UK) RN Nuclear Submarine Project placed in Class II of defence projects
1953, 30 Mar. (US) US Navy's Submarine Thermal Reactor (STR 1) went critical first time at Arco in the Idaho desert
1954, 21 Jan. (US) USS *Nautilus*, first nuclear-powered submarine launched
1954, May (UK) Naval Section increased, Captain S. Harrison-Smith appointed as head
1954, 1 Aug. (UK) Formation of the United Kingdom Atomic Energy Authority taking over from Ministry of Supply
1955, 17 Jan. (US) USS *Nautilus*, sailed on maiden sea trials
1955, 8 Dec. (UK) Proposal for 15,000–20,000 SHP nuclear power plant submitted
1956, 6 Jan. (UK) Treasury approval for HMS *Dreadnought* and Dounreay Submarine Prototype authorised
1957, 7 Nov. (UK) Neptune zero energy reactor went critical first time
1958, 14 Jan. (UK) Rickover's offer to purchase USN reactor of S3W type made
1958, Jun. (UK) US informed UK would like to purchase S5W and full machinery set
1958, 3 Jul. (UK/US) Mutual Defence Agreement signed
1958, 22 Oct. (UK) First RN Nuclear Course at HMS *Collingwood*, Fareham
1959, Jan. (UK) Department of Nuclear Science & Technology formed at RNC Greenwich
1960, 21 Oct. (UK) HMS *Dreadnought* launched
1962, 12 Nov. (UK) HMS *Dreadnought*'s reactor went critical first time
1962, 12 Dec. (UK) HMS *Dreadnought* sailed on maiden sea trials
1963, Sep. (UK) Dounreay problems with primary circuit
1963, Nov. (UK/US) Westinghouse/Rolls Royce contract expired
1965, 7 Jan. (UK) PWR 1 Core A went critical first time
1968, 15 Jun. (UK) HMS *Resolution* sailed on first UK Polaris deterrent patrol
1968, Aug. (UK) PWR 1 Core B went critical first time
1974, 16 Dec. (UK) PWR 1 Core Z went critical first time

Technical Definitions

All definitions sourced from Walker, Peter M. B., ed., *Chambers Dictionary of Science and Technology* (Edinburgh, Chamber Harrap Publishers Ltd, 2000).

Austenitic Steel: Steel containing sufficient amounts of nickel, nickel and chromium or manganese to retain austenite at atmospheric temperature, e.g. austenitic stainless steel of Hadfield's manganese steel.

Burn-up: Amount of fissile material burned up as a percentage of total fissile material originally present or fuel element performance—Heat released from a given amount of fuel GW/MW per tonne.

Cermet: Ceramic articles bonded with metal. Composite materials combining the hardness and high temperature characteristics of ceramics with the mechanical properties of metal, e.g. cemented carbides and certain reactor fuels.

Criticality: State in nuclear reactor when multiplication factor for neutron flux reaches unity and an external neutron supply is no longer required to maintain power level, i.e. the chain reaction is self-sustaining.

Enriched Uranium: Uranium in which the proportion of the fissile isotope, uranium-235, has been increased above its natural abundance.

Fast Reactor: One without a moderator in which a chain reaction is maintained almost entirely by fast fission.

Hafnium (Hf): A metallic element, it occurs in minerals containing zirconium, to which it is chemically similar, but with a higher neutron capture cross section. This makes it a troublesome impurity in the zirconium alloys used as fuel cladding.

Intermediate Reactor: One designed so that the majority of fissions will be produced by the absorption of intermediate neutrons.

Nuclear Breeder: A nuclear reactor in which in each generation there is more fissionable material produced than is used up in fission.

Pile: 'Original name for a reactor made from the pile of graphite blocks which formed the moderator of the original nuclear reactor which first went critical on the 2 December 1942 in Chicago, Illinois, US'.

Reactor Poisons: 'Other elements, especially those present as fission products, also capture neutrons very effectively. Of these xenon-135 and samarium-149 capture neutrons about a million more times effectively than elements in moderators, so very little of these elements can shut down a reactor or poison it. The gas xenon-135 readily fissions to form less damaging products in a reactor running at full power but can cause considerable problems for a reactor starting up or running at lower power for a considerable time'.

Scram: General term for emergency shut down of a plant, especially of a reactor when the safety rods are automatically and rapidly inserted to stop the fission process.

Sponge: Porous metal formed by chemical reduction or decomposition process without fusion.

Thermal Reactor: One for which the fission chain reaction is propagated by thermal neutrons and therefore contains a moderator.

The Reactor Problem: With reference to Moderators, Reactor Poisons, Control and Reactivity. 'These effects and others like the proportion of neutrons which escape the core, contribute to the *reactor problem*, whose solution determines whether a design can sustain a chain reaction without risk of meltdown. They all depend not only on the fuel, the moderator and the coolant, but also on the positions and shapes of the components. Calculating the effects of various arrangements is no easy task. Much calculation and experiment were needed to determine the most suitable materials and their disposition in the early years of the nuclear age and so solve the reactor problem. That is in ensuring that just one neutron can survive to continue the chain, no more and no less.'

Zirconium (Zr): A metallic element, the principal ores are zircon ($ZrSiO_4$), which is a very common mineral of igneous rocks and concentrated in beach sands. When purified from hafnium, its low neutron absorption and its retention of mechanical properties at high temperature make it useful for the construction of nuclear reactors.

Praise for *The Development of Nuclear Propulsion in the Royal Navy, 1946–1975*

"The nuclear enterprise, and its success or failure, was fundamental to the UK, and also the USA, maintaining the edge in the Cold War under the sea. It is a devilishly complex subject, full of twists and turns, but Gareth Jones manage to tell the story in a lucid fashion. He unravels the complexities and provides deep understanding of the technological, political, industrial and key personalities. It is a ripping yarn that anyone seeking to truly comprehend the great submarine rivalry between the Soviet Union and NATO should read."

—Iain Ballantyne, *author of 'Hunter Killers' and 'The Deadly Trade' and editor of Warships: International Fleet Review*

"The development of nuclear-powered propulsion for the Royal Navy's submarines marked a vital leap forward in the capabilities of the Senior Service. Indeed, it was an essential development if the Royal Navy was to remain in the premier league of world navies in the midst of a Cold War arms race. This book, for the first time, examines and analyses the complexities, technical, personal and diplomatic, the intrigues and twists and turns, which lay at the heart of the development of the Royal Navy's Nuclear Propulsion Programme. It was a human story as well as a technical one, that sheds new light on Anglo-American defence relations in the post-war period. For naval and diplomatic historians, this is a good book and an important one."

—Dr Harry Bennet, *Associate Professor of History, University of Plymouth*

"Taff has opened wide an important window into a notoriously closed subject. Naval history and dogged research at its best."

—Dr Sam Willis, *Broadcaster, Author and editor of Navy Records Online*

"The post-war advent of 'nuclear' was a universal game-changer. For the Royal Navy, it led to the development of a nuclear-powered propulsion programme which transformed our capability and ambition. It enabled the Submarine Service to operate almost at will in the harshest of environments for long periods of time, and to compete against a foe blessed with mass. That we prevailed in the Cold War is down to professionalism and technical superiority. It was nuclear propulsion that enabled success. Whilst this important book should appear on every "Back Aftie's" book-shelf, it will also hold a great appeal to those 'passengers' in nuclear submarines – including the author and me - who spent many years of their lives submerged but 'forward of the tunnel."

—Commodore Mike Walliker CBE, *former Submarine Commanding Officer, HMS Tireless and HMS Astute*

Contents

List of Figures

CHAPTER 1

Introduction

With the delivery of the *Frisch-Peierls Memorandum* to the British Government in March 1940 and the subsequent formation of the Tube Alloys Project, later subsumed into the Manhattan Project, the Allies' focus during World War II was on the design and construction of an atomic bomb. There was great concern that Nazi Germany was developing its own atomic weapons programme and the Allies own efforts were paramount to winning the war. However, because of the demands of World War II and the success of the Manhattan Project, it has long been forgotten that one of the first applications envisaged for nuclear fission was submarine propulsion. In the immediate aftermath of World War II nuclear-submarine propulsion programmes were instigated in the US, the Soviet Union and, here, in the UK.

On 2 December 1942, in an abandoned racket court under the stands of Stagg Field sports ground at the University of Chicago, Illinois, the first controlled self-sustaining nuclear chain reaction occurred in a reactor designated CP-1 (Chicago Pile 1). Subsequent early nuclear reactors (originally known as atomic piles) were developed to produce plutonium for atomic bombs, radio-isotopes for research and medical purposes, and for nuclear physics research. None of these reactors utilised the heat generated as a source of power. Calder Hall in the UK is generally

G. M. Jones, *The Development of Nuclear Propulsion in the Royal Navy, 1946–1975*, Security, Conflict and Cooperation in the Contemporary World,
https://doi.org/10.1007/978-3-031-05129-6_1

acknowledged as the first reactor in the world to produce power for commercial purposes in October 1956; however, its primary purpose was to produce plutonium for Britain's atomic bombs. The first purpose-built commercial reactor to produce power for electricity was the pressurised water reactor constructed at Shippingport, Pennsylvania, which went critical in December 1957. Nevertheless, the first nuclear reactor in the world to produce power that could be utilised was a prototype pressurised water reactor built at Arco in the Idaho Desert, which initially went critical in March 1953 and, after further testing, produced steam to a turbine in May. This event paved the way for a production model to be fitted into a submarine under construction at the General Dynamics Electric Boat yard at Groton, Connecticut.

The pressurised water reactor uses enriched uranium to produce heat, the water is pressurised to stop it boiling. Pumps circulate the heated water through the reactor core from where it passes through a heat exchanger in a continuous loop; the heat exchanger produces steam which is supplied to turbo-generators to produce electricity and, in the case of submarines, turbines to propel the vessel. Once the steam has condensed, it is extracted and fed back to the heat exchanger to repeat the process in a closed cycle. Although the engineering hurdles were formidable, the pressurised water reactor was selected as the best model to pursue due its compact design, simplicity of operation and the safety characteristic of its negative temperature coefficient, a description of which is given in Chap. 2. See also Fig. 1.1 for an outline of a basic pressurised water reactor propulsion plant.

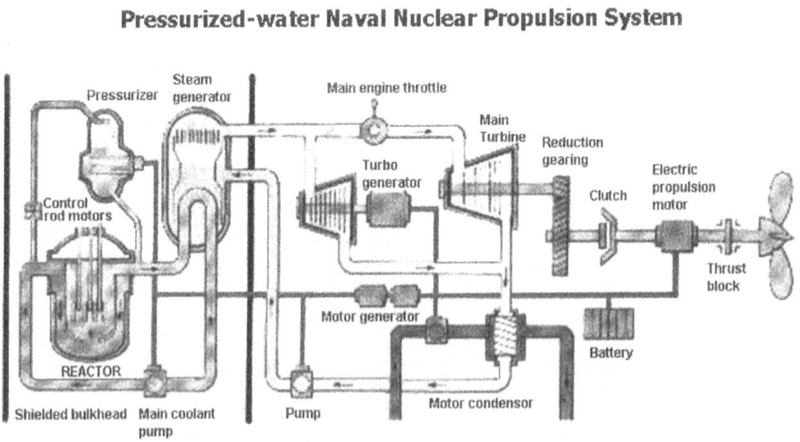

Fig. 1.1 Pressurised-water naval nuclear propulsion system

The world's first nuclear-powered submarine, the USS *Nautilus*, was launched 21 January 1954 by Mamie Eisenhower, wife of the then US president. The following year, 17 January, the USS *Nautilus* slipped down the River Thames leaving Groton, Connecticut, for her maiden sea trials. At 11:33, her Commanding Officer, Commander Eugene P. Wilkinson USN, sent the famous signal: 'Underway on nuclear power'.[1] In the intervening sixty-five years, only five other nations with the economic means have developed the technical capability to undertake the construction, operation and maintenance of nuclear-powered submarines: Russia (*Leninsky Komsomol*, SSN launched August 1957), Britain (*Dreadnought*, SSN launched October 1960), France (*Le Redoutable*, SSBN launched March 1967), China (*Long March 1*, SSN reportedly launched 1970) and India (*Arihant*, SSBN launched July 2009).[2] To gain experience of nuclear-powered submarines the Indian Navy leased a Soviet *Charlie* SSGN class submarine between 1987 and 1991, in commission with the Indian Navy it was named *Chakra*. As an integral part of a balanced blue-water navy, the nuclear-powered submarine enables these countries to project their political and military power in ways not possible by conventional means. To date, Britain has built twenty-four nuclear-powered submarines (SSN) and eight nuclear-powered and nuclear-armed submarines (SSBN) that carry the nation's deterrent.[3] Currently, the Royal Navy has three *Astute* class SSNs and four new *Dreadnought* class SSBNs either in build or on order. The first Royal Navy nuclear-powered submarine, HMS *Dreadnought*, was powered by a nuclear propulsion plant purchased from the US. Royal Navy submarines since *Dreadnought* have been powered by nuclear propulsion plants designed and built by Rolls-Royce and Associates at their Raynesway plant at Derby.[4] Since 15 January 1999, the company has been known as Rolls-Royce Marine Power Operations Ltd.

Sir Leonard Owen, who, in 1954 on the formation of the United Kingdom Atomic Energy Authority's (UKAEA), was appointed as its first Director of Engineering, wrote that at the time of setting up Britain's nuclear organisation: '...there were few scientists or engineers in Britain who were familiar with atomic energy'. Indeed, Owen noted that of the

[1] Norman Polmar and Thomas B. Allen, *Rickover* (New York, Simon and Schuster, 1982), p. 165.

[2] https://en.wikipedia.org/wiki/list_of_nuclear_submarines [Accessed 15 January 2021].

[3] NATO Acronyms: Ship Submersible Nuclear (SSN) and Ship Submersible Ballistic Nuclear (SSBN).

[4] For further details of nuclear-powered Royal Navy submarines, see Appendix.

twelve personnel starting at the UKAEA's Industrial Group at Risley, Lancashire, '...only one person knew anything about atomic energy. He was Dennis Ginns, an engineer who was home on sick leave from Chalk River'.[5] In private industry as well as the Royal Navy, the number of scientists and engineers working on nuclear matters was likely to have been negligible. The initial thrust of research and development in nuclear power was directed towards its civil application in support of the development of the atomic bomb and, later, electrical generation for the national grid. Yet it is a reflection of the political will on both sides of the Parliamentary divide, and of the scientific and engineering prowess that Britain was (is) capable of, that from these beginnings within the space of fifteen years Britain was in a position to launch its first nuclear-powered submarine, HMS *Dreadnought*. The primary aim of this book is to research, investigate and analyse the introduction of nuclear propulsion into the Royal Navy's submarine fleet, because arguably, it is part of the legacy of those political, naval and engineering decisions, made over sixty years ago, that allow Britain to "punch above its weight" on the world stage long after its Empire has ceased to exist (Image 1.1).

Overview

With economic decline setting in after World War II in Britain, and the acceptance of the right of an indigenous population to self-determination, there was a growing realisation in government that the Empire was morally, and financially, untenable. The granting of independence to the former colonies meant that there was no longer a requirement for Britain to maintain the expensive naval bases, air stations and army garrisons that were needed to defend and police these overseas territories. Many of these military assets were handed over to the new governing powers for the use of their fledgling services. Arguably, successive British Governments wanted to maintain some presence in the regions they vacated, partly in order to influence the democratic governance of their former colonies. Mainly, however, the British Government's objective was to prevent Soviet influence from filling the political vacuum of their departure and gaining

[5] Leonard Owen, 'Nuclear Engineering in the United Kingdom – the First Ten Years', *Journal of British Nuclear Energy*, (Jan., 1963), 23–32 (p. 23). Note: Chalk River, Canada, was the site of the Tube Alloys Project, the UK project to develop an atomic bomb, it later became part of the Manhattan Project.

Image 1.1 HMS/m *Dreadnought* entering Gibraltar

a presence in these regions. This was especially so in areas where major trade routes passed geographical choke-points such as the Straits of Hormuz into the Persian Gulf and the Bab El Mandeb Straits into the Red Sea. Economically and politically, it can be argued that the easiest means of maintaining a presence in a foreign region is through naval power. Unlike an air base or an army garrison stationed in a foreign country, the warship is sovereign territory and it is manoeuvrable. Diplomatic clearance from a foreign government is not required to utilise this asset and it can be positioned in areas to react to situations where it may be of greatest strategic influence. Unlike the conventionally powered surface warship, the nuclear-powered submarine requires no fuelling facilities when deployed, so it does not need to call into port or to have a Royal Fleet Auxiliary deployed with it. The nuclear-powered submarine can operate autonomously from other military and political considerations that would constrain a surface warship, which makes it a very potent unit to have at a government's disposal.

In the period after World War II up to the present day, British political and military influence has been maintained in some areas by a visible maritime presence. Initially this was achieved by aircraft carriers in the Mediterranean and the Far East and by dedicated warships in other areas deemed of importance to governments, such as the *Beira* Patrol which, between 1966 and 1976, enforced UN sanctions, in particular the oil embargo against Southern Rhodesia, at an estimated cost of some £100 million.[6] More recently, warships have been given designated patrol areas to project British influence, for instance, the *Armilla* Patrol which was instigated at the start of the Iran/Iraq war in 1980 to ensure safe passage of merchant shipping through the Straits of Hormuz. From the late 1960s however, although Britain has struggled to maintain its political and military influence, and its maritime presence has become less visible. This became more apparent in the late 1960s and early 1970s with the British Government focussed on joining the European Economic Community resulting in descending importance of the Commonwealth and Britain's worldwide authority. Britain does, however, continue to support NATO and maintain its seat as a permanent member of the UN Security Council. Significantly, Britain provides substantial submarine-focussed support to its NATO allies, the most explicit example being the nuclear deterrent patrols of the Polaris/Trident armed submarines. Also of note was the Callaghan Government's decision in the late 1970s to despatch HMS *Dreadnought* to the South Atlantic to deter possible Argentine aggression towards the Falkland Islands.[7] To this day, the nuclear-powered submarine allows Britain to maintain and exercise its political and military role on the world stage.

A typical World War II submarine, such as the German Type VIIC, displaced around 870 tons submerged, while many modern nuclear-powered submarines displace more than 5000 tons submerged.[8] The nuclear power propulsion plant has not only increased the sustained speed of the modern-day submarine, it has increased its endurance. Nuclear

[6] William Minter and Elizabeth Schmidt, 'When Sanctions Worked: The Case of Rhodesia re-examined', *African Affairs*, Vol. 87, No. 347, (Apr., 1988), 207–37 (p. 216).

[7] http://www.theguardian.com/uk/2005/jun/01/argentina.military [Accessed 3 December 2015].

[8] J. F. Starks, 'German "U"-Boat Design and Production', *Transactions of the Institution of Naval Architects*, (1948), 291–315 (p. 298).

power produces a greater electrical generation capacity which allows for more equipment to be fitted into the nuclear-propelled submarine. Because these submarines are much larger than their conventionally powered counterparts, there is more space to carry spares and stores. The type of weapons embarked have also changed; conventional torpedoes are carried alongside anti-ship missiles, cruise missiles and ballistic missiles. As such, the choice of weapons has altered the style of operations that these submarines can engage in. The Royal Navy has fired conventional munitions from its nuclear-powered submarines in support of government foreign policy in traditional operations such as the Falklands Conflict, “Operation Corporate” in 1982. This was the first occasion, and is still the only instance, of a nuclear-powered submarine launching torpedoes against a warship since World War II.[9] Increasingly, however, the target of these submarines is no longer the warship but government defence and communications infrastructure on land; as such the British Government has used Royal Navy nuclear-powered submarines to launch cruise missiles against targets in, amongst other places, Afghanistan in 2001 and, more recently, in Libya in 2011.

The period 1946–1975 has been chosen as the research time frame as the year 1946 coincides with the first employment of Royal Naval engineers and Admiralty scientists of the Royal Naval Scientific Service (RNSS) at the Atomic Energy Research Establishment (AERE), Harwell, on the problems of applying nuclear fission to submarine propulsion. The end date of 1975 captures the final core improvement of the first generation of nuclear reactors which went critical at Dounreay, Scotland, 16 December 1974.[10] Although a good source for narrative dates, Lambert’s book is not intended as a reference book, there is no bibliography and the material contained therein is sometimes vague. The period encapsulates as much of a seismic shift in the propulsion of warships as that experienced in the Victorian Navy of the 1840s which saw the introduction of steam power. Steam-driven warships were significantly more manoeuvrable than their sail counterparts and gave Royal Navy ships a greater degree of flexibility in their tactics and employment. The development, introduction and employment of steam-powered warships into the Royal Navy has been covered by many historians in a variety of media, this book will add to the

[9] HMS *Conqueror* sank the Argentine cruiser ARA *General Belgrano* 2 May 1982.

[10] Harry Lambert, ed., *Rolls-Royce: the Nuclear Power Connection* (Rolls-Royce PLC, 2009), p. 64.

already large canon of literature on the subject matter by introducing nuclear propulsion which is missing from the historiography.[11] The conventionally powered submarine is limited by the capacity of its battery for submerged speed and endurance. This is no longer the case for the nuclear-powered submarine; as such nuclear propulsion has totally changed the tactics and has enhanced the employment of the submarine driven by this technology.

At this juncture it is appropriate to illustrate why nuclear power, despite legitimate concerns about its radioactive legacy, is deemed so important as a fuel. Many people are familiar with Albert Einstein's formula, $E = mc^2$, although outside the cadre of physicists and the nuclear industry few people would do the maths to realise what this means. I would suggest that, when confronted with Einstein's formula many people would imagine an atomic explosion as an illustration of the energy released. However, the amount of energy available using Einstein's formula is made glaringly apparent when it is put into numbers. Henry D. Smyth, the author of the official history of the Manhattan Project wrote: 'It shows that one kilogram (2.2 pounds) of matter, if converted entirely into energy, would give 25 billion kilowatt hours of energy [...] Compare this fantastic figure with the 8.5 kilowatt hours of energy which may be produced by burning an equal amount of coal'.[12] To further illustrate its effectiveness as a fuel, one kilogram of matter (highly enriched uranium) is roughly the size of a pencil and one kilogram of coal is about the size of a bag of sugar; that would require approximately 2,940,000 tonnes of coal to give an equivalent power output. Smyth also points out that the 25 billion kilowatt hours of energy equated to the total energy output of the power stations of the US for two months, at 1939 figures. In today's energy-hungry society, it is worth noting that in 2019, the UK's total energy consumption was 295.48 terawatt hours.[13] It is evident that utilising this energy source in a submarine would increase its capabilities far beyond its conventionally powered counterpart, not only for speed and endurance but in its electrical power generation. Distillation plants, air purification equipment and a modern

[11] For further details on the introduction of steam power into the Royal Navy see David Evans, *Building the Steam Navy – Dockyards, Technology and the Creation of the Victorian Battle Fleet 1830–1906* (London, Conway Maritime Press, 2004).

[12] Henry D. Smyth, *Atomic energy for military purposes: the official report of the atomic bomb under the auspices of the United States government, 1945* (Princeton, Princeton University Press, 1945), p. 2.

[13] http://www.gov.uk/government/statistics/electricity-chapter-5

galley add to the quality and "comfort" of life onboard the modern nuclear-powered submarine; although, it is worth noting, that living conditions are still cramped onboard and "hot bunking" continues on some Royal Navy submarines.[14] Nevertheless, despite the lack of accommodation, its greater sensor capabilities, weapons fit and enhanced computing and communications systems equate to a more powerful adversary to be countered. The nuclear-powered submarine is one of the most potent military symbols of modern-day political power projection; it is today's capital ship.

Chapters

The next chapter, "Improving the Submersible", will investigate the developing Soviet submarine threat and the Admiralty's response to it by the pursuit of greater underwater speed and endurance for Britain's submarine fleet with research into Fast Battery Drive and air-independent engines; Malcolm Llewellyn-Jones has written on Britain's response during World War II with the modification and trials of HMS *Seraph*.[15] On this subject, Britain, and other nations, such as the US and the Soviet Union, were greatly influenced by the engineering advances and experience gained by Nazi Germany's U-boats, especially the Type XXI. Although over a hundred of these submarines were constructed, they were produced late in the war and did not see active service. However, it was accepted that the Type XXI's streamlined hull, the incorporation of the schnorkel mast and greater battery endurance giving increased underwater speed would have posed a formidable threat to the Allies had the war continued a little longer. The other submarine of great interest to the Allies was the Type XVIIB; these submarines were fitted with *Walter* turbine engines which were powered by High-Test Peroxide, H_2O_2 (HTP). In the *Walter* turbine engine, the HTP produced the required oxygen, making the engines independent of air and when dived gave a great increase in speed: '…the Germans claimed a top submerged speed of 25 knots, with two turbines'.[16]

[14] Hot bunking refers to the practice of two men sharing the same bunk in opposite watches.

[15] Malcolm Llewellyn-Jones, 'Trials with HM Submarine Seraph and British Preparations to Defeat the Type XXI U-Boat, September-October 1944', *The Mariner's Mirror*, Vol. 86, No. 4, (Nov., 2000), 434–51.

[16] Starks, 'German "U"-Boat Design and Production', p. 297.

Needless to add that this increase in speed was only sustainable so long as the HTP lasted.

Apart from research into conventionally powered air independent engines, the first chapter will also examine what was known of nuclear power as a source for motive power in the public sphere. Immediately following World War II, the general public knew of the destructive power of nuclear energy but they were less conscious of the peaceful potential uses of nuclear energy. However, some scientists, engineers and politicians were aware of the potential benefits; the peaceful use of nuclear energy was first outlined in the MAUD Committee Report of 15 July 1941.[17] Part one of the Report focussed on a rapid release of nuclear energy, the possibility of an atomic bomb; part two looked at ways of controlling the release of this energy in order to create a heat source which could then be exploited for electrical generation and motive power.[18] The chapter will examine the immediate effects of the US McMahon Act on Admiralty research into nuclear propulsion; the Quebec Agreement had been signed in 1943 and the UK had expected nuclear information and technology would continue to be exchanged with the US after the war had finished. The Admiralty was aware of the US Navy's interest in nuclear propulsion and since 1946 had maintained a small team of officers from the Royal Navy and the RNSS at the AERE, Harwell, with a brief to gain information on solving the various complex problems associated with the control and use of nuclear fission.[19]

The third chapter, "The Nuclear Option", will examine the preliminary research into utilising nuclear power for submarine propulsion. The AERE at Harwell was founded in November 1945 and Sir John Cockcroft arrived as Director in January 1946. The remit of the establishment was to: 'carry out fundamental research and development in atomic energy'.[20] Research at the AERE, Harwell, was, from the outset, primarily focussed on the civil nuclear programme which was required to support the project to develop the atomic bomb. This chapter will investigate how the Admiralty formed a cadre of nuclear experienced scientists and how the Naval Section

[17] Acronym, MAUD—Military Application of Uranium Detonation.

[18] TNA, AB 4/1014, Report by the MAUD Committee on the use of Uranium as a source of power, 1941.

[19] TNA, AB 6/81, Liaison with the Admiralty, 1946–48, Letter, R&D.240, 19 November 1946.

[20] Anon, 'Atomic Energy Research Establishment, Progress at Harwell', *British Medical Journal*, (31 July, 1948), 263–65 (p. 263).

at Harwell and the Atomic Energy Division of the Ministry of Supply went about the initial task of designing a functional power reactor for submarine use. The Royal Navy engineers and Admiralty scientists based at Harwell set about the task and had produced sketch designs within two years.[21] The chapter will also investigate the problem of reactor size stemming from the use of slightly enriched uranium and the challenge of fitting it into the envelope of a submarine pressure hull. The chapter will also consider the decisions to discontinue research into HTP and to fully focus on nuclear propulsion from the mid-1950s onwards.

Research in the fourth chapter, "The Pressurised Water Reactor" (PWR), will examine the challenge of building a nuclear reactor and the engineering and quality assurance difficulties that needed to be overcome to harness the power of nuclear fission. A typical pressurised water reactor power plant, as used in submarine propulsion, is composed of several components which are collectively known as the primary system, these will be detailed in the chapter. Due to the radiological problems involved, it is essential to select the necessary appropriate materials and to stipulate quality specifications to allow the manufacture of primary system components to the highest engineering standards practicable. The machinery and equipment need to be as robust and reliable as possible as failure of any part could have catastrophic consequences for the reactor plant and, indeed, the submarine. Maintenance and repair of any component in the primary system cannot be undertaken whilst the reactor is critical, and even once the reactor plant is shut down, entry into the reactor compartment would not be possible for a period of time afterwards. In considering nuclear physics, metallurgical problems such as finding materials that could withstand the intense heat, pressure and irradiation of the reactor vessel had to be overcome. Materials chosen would also have to have a low neutron capture cross section to allow fission to take place. Attention will be given to the selection of uranium/zirconium fuel elements over uranium oxide/steel fuel elements, which was to have a major significance for the nuclear submarine programme. Conversely, a material also had to be found that would readily absorb neutrons to allow the nuclear reaction and the power available to be controlled and shut down. All these materials would have to be chosen carefully so they do not to chemically react

[21] Eric J. Grove, *Vanguard to Trident: British Naval Policy Since World War II* (London, The Bodley Head, 1987), p. 230.

with each other, and would require testing in a specially designed zero energy reactor as part of the reactor physics for calculating criticality.

With these factors in mind, the engineering problems to be solved by staff at Harwell were many and diverse; however, it would be industry that would have to supply the materials to meet the composition and qualities required by the designers. An effect of demanding the highest engineering standards was that quality assurance was to become the foremost means of managing suppliers by introducing exacting standard requirements for material supplied to the Admiralty and of testing and controlling those standards. Research will also consider what information the Americans were publishing in peer-reviewed articles that could be used by the British in their quest for a naval reactor. By the mid-1950s the US Navy was producing handbooks and papers on reactor engineering as a means of supporting industrial research into the subject.[22]

The fifth chapter, "HMS/m *Dreadnought*", will investigate the influence of Admiral Hyman G. Rickover USN and his relationship with Admiral of the Fleet, Lord Louis Mountbatten, which was pivotal in Britain not only by enabling a leap forward in reactor engineering knowledge, but also, more importantly, through acquiring US methods for engineering acceptance. The chapter will explore the offers to purchase an American submarine reactor and the consequences for the UK-designed PWR 1; this chapter will also consider how far the Admiralty's own programme had progressed when the US made its offer to sell a submarine propulsion reactor. During research it has been noted that there are a number of anomalies in Philip Ziegler's biography of Mountbatten concerning the purchase of the S5W plant and machinery, which other authors have quoted, and these will be addressed for future reference. The chapter will end by investigating the consequences of delays in the Admiralty's nuclear programme, the signing of the Mutual Defence Agreement and the subsequent purchase of an American reactor and its associated machinery for the Royal Navy's first nuclear-powered submarine.

Chapter 6, "Nuclear Training and Dounreay", will examine what steps were taken by the Admiralty to establish a cadre of qualified engineers and mechanics for watch-keeping purposes prior to launching HMS *Dreadnought*. Research will examine the nature, scope and periodicity of training and how it was introduced into the Royal Navy so that its engineers and mechanics were able to operate the nuclear power plant and its

[22] Richard G. Hewlett & Francis Duncan, *Nuclear Navy 1946–1962* (Chicago & London, The University of Chicago Press, 1974), p. 139.

associated machinery. Research will investigate Dounreay's role as an important practical training facility and its primary purpose as a prototype for the whole submarine primary and secondary machinery and for testing new types of core and establishing maintenance techniques and operating procedures. The chapter will explore the initial design and construction of Britain's first naval reactor (PWR 1) and the setting up of the Dounreay Submarine Prototype (DS/MP) in Caithness on the north coast of Scotland.[23] It was essential to ensure that the nuclear power plant design would fit into a submarine pressure hull. It was also deemed necessary to undertake trials of the plant and machinery as well as gaining operating experience prior to fitting into the first submarine as indeed, a major problem did come to light during commissioning trials. This was the same procedure the Americans had used when they built their prototype, the Submarine Thermal Reactor (STR 1) in the Idaho desert in 1953 prior to fitting a production reactor into the USS *Nautilus*. Any engineering problems found during trials on the prototype could be rectified before fitting the production model into the submarine hull.[24]

The final chapter, "Future Developments", will look at how advantage was taken of the knowledge acquired from the US and managed to develop new generations of reactor cores for new classes of submarines and which could also be back-fitted during refits into the submarines with older types of cores. The chapter will look at the relentless Treasury pressure on the Admiralty to save money; the Admiralty made references in their correspondence with the Treasury to requests for further information from the US authorities in the knowledge that it would not be forthcoming. By the end of 1962, the Treasury had stopped asking the Admiralty to make requests to the US. The chapter will also consider other avenues of research that the Admiralty and Rolls-Royce and Associates undertook; not only were they looking at improving the core of the pressurised water reactor, other types of reactors were also being investigated for their potential application in submarines. The chapter will conclude by examining the development and improvement of the associated secondary machinery systems, and the all-important advances in air purification without which the benefits of nuclear propulsion could not have been fully realised.

[23] DS/MP referred to the reactor, the establishment was known as the Admiralty Reactor Test Establishment (ARTE) and later as HMS *Vulcan*. For clarity, the name Dounreay will be used throughout to refer to the reactor and the establishment.

[24] Hewlett & Duncan, *Nuclear Navy*, pp. 164–165.

Literature Review

Submarines have long held a fascination in the public's imagination; this can be seen in the number of books and articles on the subject matter. All interests are catered for, from fiction to fact, as illustrated by Jules Verne's classic, *20,000 Leagues Under the Sea* (1873). In 2015, Peter Hennessy and James Jinks had their book published, *The Silent Deep: The Royal Navy Submarine Service since 1945*, which relates the history of the Royal Navy Submarine Service from the opening days of the Cold War to modern times and includes personal accounts. In film too, the submarine features in popular culture, from the many gritty World War II dramas such as *Das Boot* (1981), through the numerous films of the James Bond franchise to productions featuring the Cold War, such as *The Hunt for Red October* (1990). The main theme of the vast canon of submarine literature is concentrated on the "cat and mouse" tactics of the hunter and the hunted. They feature famous characters, both factual and fictional, and they cater to the public's interest in the exploits and tactics of underwater combat and the missions undertaken by these secretive leviathans of the deep.

There are, additionally, many learned articles on submarines to be found in journals such as *Mariner's Mirror* and *International Security*.[25] Many of these concentrate on the historical significance of submarine development and activity during World War I and II. Nevertheless, articles can be found relating to the post-war era which relate to general details of the nuclear-powered submarine, but the one area that is almost always overlooked, if not neglected, is the submarine's propulsion system and this is especially so in the era of nuclear propulsion. An extensive search for literature on the history of nuclear power reactor plant development in the Royal Navy has reaped a very limited harvest. Much of what has been written on the post-war history of Anglo-American nuclear relations has focussed on Britain's attempts to develop the atomic and hydrogen bombs without assistance from its former ally, America. Limitations on the transfer of nuclear technology and information from the US to a third party that could be used in the development of an atomic bomb, or other military use, was imposed by the American Atomic Energy Act of 1946, commonly referred to as the McMahon Act.[26]

[25] See for example Karl Lautenschlager, 'The Submarine in Naval Warfare, 1901–2001', *International Security*, Vol. 11, No. 3, (Winter, 1986–1987), 94–140.

[26] Michael Goodman, 'With a little help from my friends: The Anglo-American Atomic Intelligence Partnership, 1945–1958', *Diplomacy and Statecraft*, Vol. 18, No. 1, (Mar., 2007), 155–83 (p. 155).

America was keen to keep the monopoly on nuclear power to itself for as long as was possible. This went to the extent of refusing to cooperate further with Britain, its closest wartime ally and partner in the Manhattan Project whose crucial contribution has been described as: '…so small in terms of men and resources, so large in terms of its consequences'.[27] Authors who have written on the subject of Anglo-American nuclear relations include John Baylis and C. J. Bartlett. In one article, Baylis argues that the keystone to this relationship: '…was the desire on both sides of the Atlantic for an interdependent nuclear relationship'.[28] The Mutual Defence Agreement Act of 1958 is central to this "special" relationship and played a critical role in the development of Britain's first nuclear-powered submarine, HMS *Dreadnought*.[29]

The history of the US Navy's involvement in the development of nuclear submarine propulsion under the leadership of Admiral Hyman G. Rickover USN was written less than twenty years after the USS *Nautilus* sailed for sea trials. Richard G. Hewlett and Francis Duncan's seminal work, *Nuclear Navy 1946–1962*, captures the early history of the US Navy's attempts to overcome the engineering problems associated with manufacturing a nuclear reactor compact enough to fit into a submarine's hull. In the UK, three important lectures have been delivered on the Royal Navy's involvement with nuclear propulsion, these were later published as articles. The first lecture, the 54th Thomas Lowe Gray Memorial Lecture, was delivered by Vice Admiral Sir Ted Horlick, who at that time was the Royal Navy's Engineer-in-Chief, to the Institution of Mechanical Engineers, 26 January 1982. Horlick gives a summary of the Royal Navy's involvement with nuclear propulsion from the creation of the Admiralty Development Establishment Barrow, to the development of the second generation of the Royal Navy's pressurised water reactors (PWR 2).[30] The

[27] Margaret Gowing, 'Britain, America and the Bomb', in *Retreat from Power: Studies in Britain's Foreign Policy of the Twentieth Century, Volume Two After 1939* (London, Macmillan, 1981), p. 121.

[28] John Baylis, 'The 1958 Anglo-American Mutual Defence Agreement: The search for Nuclear Interdependence', *Journal of Strategic Studies*, Vol. 31, No. 3, (Jun., 2008), 425–66 (p. 426).

[29] Treaty Series No. 41 (1958) Agreement between the Government of the United Kingdom of Great Britain and Northern Ireland and the Government of the United States of America for Co-operation on the Uses of Atomic Energy for Mutual Defence Purposes Cmnd.537. Hereinafter referred to as the Mutual Defence Agreement.

[30] Ted Horlick, 'Submarine Propulsion in the Royal Navy', *Journal of Naval Engineering*, Vol. 27, Book 1, (Jun., 1982), 1–22.

second lecture was delivered to the Institute of Marine Engineers, 10 October 1995, by another Engineer-in-Chief, Vice Admiral Sir Robert Hill. In this lecture, Hill reflects on the Royal Navy's achievements of forty years' experience with nuclear propulsion.[31] The third lecture was also delivered by Vice Admiral Hill, on this occasion to the Institution of Mechanical Engineers, 19 April 2005. The subject of the 77th Thomas Lowe Gray Memorial Lecture was Admiral Hyman G. Rickover USN and the UK Nuclear Submarine Propulsion Programme. This lecture detailed the legacy of Rickover's influence over the Royal Navy's nuclear propulsion programme, the impact of the Mutual Defence Agreement and the consequent purchase of the S5W nuclear reactor and associated propulsion machinery, which was later fitted into the Royal Navy's first nuclear submarine, HMS *Dreadnought*.[32] Vice Admiral Hill continued to deliver this lecture until very recently to Royal Naval engineers and mechanics at the Royal Navy's Marine and Air Engineering School, HMS *Sultan*, based at Gosport, Hampshire.[33] Because of security constraints at the time these lectures were delivered, there was only a limited amount of material that could be used and the histories are, by necessity, very much a broad sweep of nuclear propulsion in the Royal Navy.

Nuclear Historiography

The writing of nuclear history as a subject is by its secretive, some would possibly argue controversial, nature in its infancy in the UK. There are three writers who stand out among their contemporaries, all of whom were women. Uniquely, all three were "insiders", employed by the establishments and the authorities they wrote about. The Grigg's Committee report in 1954 on Departmental Records resulted in the Public Records Act being passed in 1958. The following year, Margaret Gowing was employed by the United Kingdom Atomic Energy Authority after they sought to appoint a historian and archivist to compile their numerous

[31] Robert Hill, 'Forty Years On: The UK Naval Nuclear Propulsion Programme', Address to the Institute of Marine Engineers, 10 October 1995.

[32] Acronym US Naval Reactor: S5W refers to Submarine reactor, 5th generation, Westinghouse build.

[33] Robert Hill, 'Admiral Hyman G Rickover USN and the UK Nuclear Submarine Propulsion Programme', *International Journal of Naval History*, Vol. 4, Iss. 2, (Aug., 2005), 1–16.

records which were dispersed over several sites across Britain.[34] By 1964, Gowing had produced the first of three seminal works on the subject of Britain's association with nuclear power, *Britain and Atomic Energy, 1939–1945*. Ten years later, Gowing was assisted by Lorna Arnold, which resulted in another book in two volumes being published; *Independence and Deterrence: Britain and Atomic Energy 1945–52: Volume 1 Policy Making*, and *Independence and Deterrence: Britain and Atomic Energy 1945–52: Volume 2 Policy Execution*.

These volumes constitute the official history of the British Government's development of the atomic bomb and civil nuclear power plant building programme. The civil programme was developed to support the production of enriched uranium and, through the civil nuclear power plants, plutonium; these materials were required for the construction of the atomic bomb. Gowing's focus is on both of these nuclear projects; however, Gowing does comment on what was happening at the AERE, Harwell, during the early years in relation to the nuclear submarine propulsion programme. Gowing noted that in September 1950 the UKAEA's research and development staff at Harwell had a high priority for a submarine reactor using enriched uranium.[35] A design study was awarded to the firm Metropolitan-Vickers; however, the reactor design was deemed too large to fit in a submarine hull and was discontinued.[36] Gowing noted that later in 1951 the submarine nuclear reactor project was given a lower priority by the Admiralty because of the expense of having to use enriched uranium. It could be argued that it was not the expense of using enriched uranium but the lack of it and the government's priority to manufacture the atomic bomb that led to the decision to give a higher priority to the fast breeder reactor programme which would produce the materials required for the atomic bomb.[37] The third woman is Katherine Pyne, who was official historian at the Atomic Weapons Establishment, Aldermaston.

[34] ODNB, Robert Fox, 'Gowing, Margaret Mary (1921–1998)', *Oxford Dictionary of National Biography*, Oxford University Press, 2004. http://www.oxforddnb.com/view/article/71257 [Accessed 21 October 2015].

[35] Natural uranium is ^{238}U and 0.7% of this consists of the isotope ^{235}U. It is this isotope that more readily releases its neutrons during the fission process. Enriched uranium involves processing natural uranium to contain a greater percentage of ^{235}U.

[36] Margaret Gowing, *Independence and Deterrence: Britain and Atomic Energy 1945–52: Volume 1, Policy Making* (London, Macmillan Press, 1974), p. 239.

[37] Gowing, *Independence and Deterrence, Volume. 1*, p. 445.

Pyne contributed to the work of Lorna Arnold, *Britain and the H-Bomb* (2001), and Peter Hennessy, *Cabinets and the Bomb* (2007).

The focus given to the McMahon Act by historians begs the question why the nuclear propulsion of submarines element of Britain's post-war ambitions has been neglected. The most probable reason is a combination of reluctance on the part of the Ministry of Defence, the Royal Navy and the nuclear industry in general to engage with historians. This reticence to divulge information on nuclear submarine propulsion plants, however small and insignificant, is a direct result of the Mutual Defence Agreement. Article III covers the 'Transfer of Submarine Nuclear Propulsion Plant and Materials'. However, it is made clear in Article VII, which covers the dissemination of information, that communication of any information to a third party, unless by agreement, is strictly prohibited.[38] It is, perhaps, easy to understand a historian's reluctance to undertake this work without official authorisation from the government departments involved. Support for this research from the office of the First Sea Lord, at that time Admiral Sir George Zambellas KCB DSC ADC, and the office of the Director of Nuclear Propulsion, Commodore Mark Adams, has assisted in allowing privileged access to MoD files from that period yet to be released to the National Archives.

In 2000, Arnold wrote: 'Britain's official nuclear history ranged widely – covering the wartime story of Britain's crucial contribution to the Manhattan Project; ... [to] the reactor fire at Windscale, in Cumbria. The programme remains incomplete'.[39] HMS *Dreadnought* was commissioned 17 April 1963, and it is forty-one years since she was decommissioned and laid up at Rosyth Dockyard in 1980. There are a few people alive today that were actively involved during the 1950s with the research at the AERE, Harwell, and later with the Dreadnought Project Team (DPT), based at Bath. These engineers had first-hand experience in dealing with their American counterparts during the purchase of the S5W reactor plant from Westinghouse in America and the associated propulsion machinery from General Electric. Included in this group of first-generation Royal Navy nuclear engineers is the first Deputy Marine Engineering Officer (DMEO) of HMS *Dreadnought*, Rear Admiral Peter Hammersley CB OBE, who undertook sea training in the American submarine USS

[38] Mutual Defence Agreement, 1958, p. 5.

[39] Lorna Arnold, 'A letter from Oxford: The History of Nuclear History in Britain', *Minerva, A Review of Science, Learning and Policy*, Vol. 38, (2001), 201–19 (p. 201).

Skipjack between October 1959 and May 1960 prior to his appointment to HMS *Dreadnought*. [40] It is deemed important to gather statements of their witness to the early days of nuclear propulsion to add to the nuclear history of this country.[41]

The Nuclear Submarine in Context

At the end of World War II, the Royal Navy had a large fleet but many ships were obsolete and inefficient. The capital ship of the day, the battleship, was extremely vulnerable to air attack as illustrated by the Royal Navy's Fleet Air Arm attack on the Italian fleet at Taranto Harbour in November 1940, which influenced the more infamous Japanese attack on the US fleet at Pearl Harbour in December 1941. The final nail in the coffin for the battleship were the Japanese air attacks on 10 December 1941, on elements of Force Z, a Royal Navy squadron sent to deter Japanese expansion in south-east Asia. This resulted in the sinking off Malaya of HMS *Repulse* and HMS *Prince of Wales*, this sea battle should earn it the sobriquet, "the day the battleship died". The last battleship to be built for the Royal Navy, HMS *Vanguard*, was completed after the war but after a short service career she was decommissioned in 1960.[42] The concept of the cruiser was also reviewed by the Admiralty and resulted in only three new completions after the war. The foundations were laid for a more efficient fleet based around the aircraft carrier, submarine and Anti-Submarine Warfare (ASW) forces, and these improvements to the fleet began to appear a decade later.[43]

As the British, Japanese and Americans had proved during World War II, the aircraft carrier was considered the new capital ship; Britain was determined to keep at least one operational aircraft carrier in the Far East station, the Mediterranean and in Home Waters, and the Americans continued to refine their concept of the carrier strike force. In the post-war period, many World War II-era destroyers were scrapped; however, some were taken into dockyard hands and fitted out with modern machinery, weapons and enhanced ASW capabilities. However, it had been

[40] Private papers, Rear Admiral Peter Hammersley.

[41] Since completing research Rear Admiral Hammersley passed away in January 2020.

[42] Mike Critchley, *British Warships Since 1945: Part 1* (Liskeard, Maritime Books, 1980), p. 10.

[43] C. J. Bartlett, *The Long Retreat: A Short History of British Defence Policy, 1945–1970* (London, Macmillan, 1972), p. 34.

acknowledged that despite the improvements in surface ASW, the new submarines entering service with faster submerged speeds would still be extremely difficult to detect and destroy. During the 1950s, the Royal Navy was finding it harder to counter Army and Royal Air Force criticism of its expenditure on aircraft carriers; this, of course, was part of the wider political/military debate in which the Royal Navy struggled to argue an effective case for the fleet in war with a growing greater political and military reliance now placed on nuclear deterrence.[44] The Royal Navy's aircraft carriers were also deemed to be vulnerable to Soviet missile attack and the RAF strongly argued its case that they would be able to cover maritime areas from their air bases which, at that time, were positioned across the globe. The Royal Navy eventually lost the argument resulting in the cancellation of the CVA01 replacement carrier project in the mid-1960s. It is worth noting that although the carrier was "politically" dead, the Royal Navy proceeded with the concept of the "through-deck cruiser" resulting in the *Invincible* class carriers which began to commission in the early 1980s. By the mid-1960s, however, the nuclear-powered submarine was seen by many to be the capital ship of the future. The nuclear submarine was deemed to have the potential to engage in covert operations in forward areas and gather intelligence. These were operations, sometimes politically sensitive, that the Army and Royal Air Force could not undertake with the same degree of stealth and invulnerability, and this presented an opportunity to the Royal Navy to argue the case for a fleet of nuclear-powered submarines. Indeed, intelligence-gathering operations, "up North" as it is colloquially known, began in the early 1950s employing conventionally powered submarines, as using a nuclear-powered submarine would mean greater stealth and greater endurance resulting in better reconnaissance opportunities.[45]

The largest threat to Western Europe and America in the immediate post-war period was that of the Soviet submarine fleet. In one of his Quarterly Newsletters in 1957, Mountbatten noted that according to Jane's Fighting Ships there were about 400 Soviet submarines and some 100 under construction.[46] To counter the submarine threat, it was argued

[44] Laurence W. Martin, 'The Market for Strategic Ideas in Britain: The "Sandys Era"', *The American Political Science Review*, Vol. 56, No. 1, (Mar., 1962), 23–41 (p. 26).

[45] Peter Hennessey & James Jinks, *The Silent Deep: The Royal Navy Submarine Service since 1945* (London, Allen Lane, 2015), p. 94.

[46] Broadlands Archive, MB1/I300, First Sea Lord's Quarterly Newsletters, 1 November 1957.

that the only efficient way to carry out this function was to fight like with like, and from 1949 the interception and destruction of enemy submarines became the primary function of the Royal Navy's submarine fleet.[47] The development of conventionally powered submarines incorporating World War II technologies from the German Type XXI U-boat, the attempts to master *Walter* HTP propulsion during the 1950s and the eventual adoption of nuclear propulsion can be seen as part of the Admiralty's progressive attempts not only to counter the Soviet submarine threat, but also as having the means of destroying that threat should the need arise. The submarine threat continued throughout the Cold War as the Soviet Union continued to invest in submarine technologies in order to maintain not only a credible ballistic missile force but also to enhance the hunter-killer capabilities of their SSNs.

Summary

The time has come for the development of nuclear propulsion in the Royal Navy to be written and to assess the political, military and engineering decisions which led to the building, operation and deployment of these potent naval assets. Research will look at such decisions which allowed private industry to develop the reactor core for nuclear submarine propulsion rather than the state, through the UKAEA, keeping a monopoly on core design. Assessing these decisions will allow a better understanding of the relationships between the Admiralty, Royal Navy, the UKAEA staff based at Harwell and private enterprise, and how these relationships impacted upon and influenced the engineering decision-making process.

Research will investigate the role of the First Sea Lord, Earl Mountbatten, who had a Royal Naval weapons engineer appointed to the US Polaris project as early as 1955.[48] Given the struggle the Royal Navy had in defining their role in total war, it appears that Mountbatten's long-term aim for the Royal Navy was to carry the nation's deterrent. Strategic thinking at the time was moving towards sea-based deterrent forces and away from the static missile sites and free-falling bombs which were deemed vulner-

[47] TNA, ADM 1/20414, Functions of operating submarines: investigation and report into possibilities of operating submarines as convoy escorts, or with hunting groups in a fighter role, 1947–49; 1954, Report, TASW 297/49, October 1949.

[48] Broadlands Archive, MB1/J40, letter, Mountbatten to Rear Admiral Galantin USN, 21 January 1965.

able to improved anti-missile defences in target areas. For a sea-based deterrent to be totally effective, to remain undetected and an ever present threat, nuclear propulsion had to be mastered and operating experience gained so that a case could be made to government should an opportunity to purchase the Polaris system arise, as it did in due course with the Nassau Agreement in December 1962.

This book will appraise the development of nuclear submarine propulsion during the third quarter of the last century and will add a vital piece of missing literature to the naval historiography of this country. With support from the office of the Director of Nuclear Propulsion, Commodore Mark Adams RN, a Confidentiality Agreement has been signed allowing access to unreleased files. A number of retired Royal Navy engineering officers involved in the nuclear propulsion programme during the 1950s and 1960s have given freely of their time and allowed access to their private papers. This will enable personal perspectives to be added to the events of that period which continue to influence the Royal Navy of the twenty-first century.

CHAPTER 2

Improving the Submersible

The submarine was initially developed as a sea denial platform by continental countries with weaker navies in comparison to the Royal Navy. The French launched the battery-powered *Gymnote* in 1888, this vessel was armed with two torpedoes.[1] In April 1900, the US Navy purchased the *Holland*, this submarine joined an American fleet comprised of two armoured cruisers, six monitors, seven (first and second rate) battleships and seventeen other vessels.[2] In response to these innovations, the Royal Navy built five *Holland*-class submarines under licence by Vickers at Barrow-in-Furness. During both World Wars, the submarine was used to destroy the commercial trade that Britain and its allies relied upon; aircraft and destroyers were the main means of dealing with the submarine threat. It was only with the advent of nuclear propulsion that the submarine would become the ultimate hunter and killer of submarines.

During World War I, Germany began designing U-Boats with an emphasis on greater surface performance to allow improved transit times to the operational area. During this period, the majority of targets were attacked on the surface with the deck gun, submarines dived only when

[1] https://www.britannica.com/topic/gymnote [Accessed 08 November 2021].

[2] Frank Uhlig Jr., 'The Submarine, 1776–1918', *Naval War College Review*, Vol. 57, No. 2, (Spring 2004), 146–158 (p. 146).

G. M. Jones, *The Development of Nuclear Propulsion in the Royal Navy, 1946–1975*, Security, Conflict and Cooperation in the Contemporary World,
https://doi.org/10.1007/978-3-031-05129-6_2

absolutely necessary due to the limited endurance of their batteries. This was the design philosophy that the German navy, and other navies adopted and improved on during the inter-war years; however, this design philosophy would start to show its weakness with improvement in aircraft design and the adoption of new technologies. The attrition rate for German U-boats during World War II was an important lesson in understanding the vulnerability of the submarine whilst surfaced. The development of radar and its employment in long-range aircraft in finding and attacking the surfaced submarine made research into air-independent engines a primary task in the post-war period. The Germans had employed the schnorkel in an attempt to limit the losses to its submarine fleet; they had also developed U-boats with a higher battery capacity and more efficient motors. The design and use of HTP and the *Walter* turbine were also well developed by Germany and were entering production towards the end of the war. However, all these improvements could not detract from the fact that the conventionally powered submarine would still be limited by the amount of fuel it could carry and would still be little more than a submersible reliant on the requirement to surface periodically. Therefore, it is important to have a brief understanding of the discovery of nuclear fission and how this process, more commonly associated with the atomic bomb and civil nuclear power stations, became, from its discovery, associated with the problem of producing what Admiral Chester W. Nimitz USN and others called the "true submarine".[3] Within twelve months of the discovery of nuclear fission and the suggestion that a chain reaction may be possible to control, proposals were put forward in France and America for applying this new science to the problem of air-independent propulsion for submarines. A nuclear-powered propulsion plant would enable a submarine to roam the depths at will, free from the constraints imposed by the diesel engines and battery-driven motors of the conventionally powered "submersible" submarine with its requirement to operate on or near the surface. The nuclear-powered submarine would become a formidable maritime threat and a difficult obstacle to be countered by an adversary. To quantify the difference between conventionally powered submarines and their nuclear-powered counterparts: 'If the old fashioned U-boats overall submerged performance is represented by m, then the Type XXI would be 10m, the Walther boat 40m and the nuclear submarine at 5,000m'.[4]

[3] Hewlett and Duncan, *Nuclear Navy*, p. 12.

[4] Vice Admiral Sir Arthur Hezlet, *The Submarine and Sea Power* (London, Peter Davies, 1967), p. 244.

Introduction of World War II Submarine Developments

In recording developments in submarine propulsion methods, it is necessary to look at why and when the Admiralty began to investigate air-independent propulsion systems and to understand the reasons why the nuclear option was eventually chosen as the preferred method. One of the first problems to be tackled by the Admiralty was the lack of standardisation in submarine propulsion which had hampered repair and maintenance during the war. At the outbreak of World War II, the Royal Navy had a number of classes of submarine in commission, the World War I-era 'H' and 'L' classes and the 'O' and 'P' classes commissioned during the late 1920s and early 1930s. More modern submarines were also commissioned; designed and built in the mid-1930s to replace the 'H' class, production of the 'U' and 'S' classes would increase greatly during World War II. Finally, the larger 'T' class, which could carry a greater outload of twenty torpedoes, began to commission in 1938.[5] Astonishingly, there were fourteen types of diesel engines fitted in these submarine classes, in the UK there were four yards where submarines were being built; Vickers at Barrow, Cammell Laird at Birkenhead, Scotts at Greenock and Chatham Dockyard, all of them built their own engines.[6] Such a miscellany of equipment had a detrimental effect on the logistical organisation of spares for maintenance and repairs which would not be resolved until the late 1950s. Design of a prototype 16-cylinder 'V' diesel engine, the *Muscovic*, had begun in 1937; however, work stopped on the outbreak of war but development of the design specification continued.[7] In 1947, Chatham Dockyard started design work on the Admiralty Standard Range (A.S.R.1) engine incorporating many new specifications, trials and testing were conducted in late 1950 prior to the engine being accepted into service. Based on the *Muscovic*, the A.S.R.1 supercharged engine was fitted into the new generation of diesel-electric submarines, the *Porpoise* class; the engine would also be fitted in the follow-on *Oberon* class. Whilst development of

[5] Mike Critchley, *British Warships Since 1945: Part 2* (Liskeard, Maritime Books, 1981), p. 35.

[6] L. A. Taylor, 'Submarine Engineering Problems and Development 1939–1945', *Journal of Naval Engineering*, Vol. 2, Book 2, (Jul., 1948), 115–35 (p. 116).

[7] W. H. Sampson, 'A.S.R.1', *Journal of Naval Engineering*, Vol. 9, Book 1, (Jan., 1956), 56–65 (p. 58).

a standard submarine engine was proceeding, there were urgent requirements to be dealt with the other submarines in the fleet.

In the aftermath of World War II, the Royal Navy had a large number of operational 'S', 'T' and 'U' class submarines, they also had new 'A' class submarines in the process of being commissioned and built. It could be argued that the Battle of the Atlantic had been won in 1943 when German U-boat losses were running at an unsustainable rate due to improvements in, and the employment of, radar and long-range aircraft. However, British intelligence learnt that Germany was continuing to produce new submarines with higher performances in speed and underwater endurance; as such the submarine threat to allied shipping in the Atlantic remained real until the end of the war in 1945. Britain and her Allies were quick to learn from German experience, through intelligence gained, the Admiralty was made aware of the existence of a new submarine, the Type XXI, but in some respects much of what was learnt through intelligence was already known. The schnorkel mast had been fitted and trials conducted on Dutch submarines before the war. When the Netherlands was overrun in 1940, four Dutch submarines made it to Britain and the Germans captured the remainder. Thus, the technology was available to both navies.[8] The techniques employed and the parameters required to streamline a submarine to reduce drag and so increase speed with little or no increase in fuel consumption were also widely known to the Admiralty.[9] It is sufficient to note that, at that stage of the war the Royal Navy saw no requirement to incorporate the schnorkel mast into its submarine fleet; some drawings were made during 1942–1943 for the 'U' class submarines but were never completed.[10] However, a decision was later made to convert and streamline one submarine, HMS *Seraph*, in order to train surface ships against a high underwater speed target with the potential performance of Germany's Type XXI submarine.[11]

HMS *Seraph* was taken into dockyard hands 16 June 1944, one month after the first Type XXI, *U2501*, had been launched. HMS *Seraph's* gun was removed, and fairings were fitted around the masts and periscopes.

[8] Mark C. Jones, 'Give Credit Where Credit Is Due: The Dutch Role in the Development and Deployment of the Submarine Schnorkel', *The Journal of Military History*, Vol. 69, No. 4, (Oct., 2005), 987–1012.

[9] Llewellyn-Jones, *The Mariner's Mirror*, p. 435.

[10] L. A. Taylor, 'Submarine Engineering Problems and Development 1939–1945', p. 128.

[11] Malcolm Llewellyn-Jones, 'The Royal Navy on the Threshold of Modern Anti-Submarine Warfare, 1944–49', (unpublished PhD thesis, University of London, 2004), p. 23.

Shutters were also fitted to the torpedo tube openings in order to streamline her. Paul Kemp noted: '*Seraph's* underwater resistance was reduced by 45 per cent and she achieved a speed of 14.25 knots dived compared to 8.82 knots with an unconverted S boat'.[12] Comprehensive trials were conducted and two more 'S' class submarines were converted in 1945, HMS *Satyr* and HMS *Sceptre*, to assist in training the surface fleet in tactics against the Type XXI.[13] As noted previously, this class of submarine did not materialise operationally during the war, however, as Llewellyn-Jones noted: '...it re-emerged in the real, or perceived, menace from the Soviet submarines exploiting the Type XXI technology'.[14] It was this looming threat, not readily apparent in 1945–1946, that Britain faced and proved the prime mover in adopting technological improvements such as the schnorkel in its submarine fleet. By 1947, the deterioration in relations between the Western Powers and the Soviet Union was to provide the stimulus to find a means of air-independent propulsion as a method of combatting the growing Soviet submarine threat.

Towards the end of the war the schnorkel, known as the snort in the Royal Navy, was seen as the quickest means of increasing a submarine's dived endurance, and many submarines were taken into dockyard hands to have the mast and associated equipment fitted, the first being HMS *Truant* in the closing stages of the war.[15] After the war, a series of so-called snort patrols were conducted in order to assess the equipment and increase experience. The patrols also helped to gain information on any physiological problems with the crew which may be associated with being submerged for prolonged periods. These patrols can be seen as an acknowledgement that the Admiralty realised that submariners were entering into areas where medical science may be needed to counter the possible ill effects of the submarine's atmosphere and other factors experienced during an extended dived period. HMS *Taciturn* conducted the first prolonged dive in temperate waters followed by HMS *Alliance* which carried out her dived patrol in tropical conditions. HMS *Alliance* began her patrol 9 October 1947 diving off the Canary Islands and surfaced thirty days later,

[12] Paul Kemp, *The T-Class Submarine: The Classic British Design* (London, Arms and Armour Press, 1990), p. 130.

[13] Llewellyn-Jones, *The Mariner's Mirror*, p. 446.

[14] Llewellyn-Jones, *The Mariner's Mirror*, p. 447.

[15] Jones, 'Give Credit Where Credit Is Due', p. 1009.

8 November 1947 off Freetown.[16] The final experiment in this series of extended dives was conducted by HMS *Ambush* in the Arctic between Jan Mayen and Bear Island during February and March 1948.[17] The patrols gave not only a great deal of operational experience in snorting techniques, in many respects they also contributed to the foundations of improving atmosphere control and domestic arrangements that would be required so that nuclear propelled submarines could take full advantage of their increased endurance.

As part of the Allies agreement covering the distribution of war spoils, ten U-boats each were to be divided between the US, the Soviet Union and Britain, the remainder were to be disposed of. During 1946–1947, four Type VIICs, one of each Type IXC and Type XXIII, and, most important, four of the advanced Type XXI submarines were transferred to the Allied navies. However: 'The Soviets also captured fourteen incomplete Type XXIs in an advanced state of construction, together with a large number of prefabricated sections, when they entered Danzig in 1945'.[18] Although the British and Americans conspired to keep any of the Type XVIIB *Walter* submarines from falling into the hands of the Soviet Union, the Soviets did gain access to the technology. At Blankenberg, the Soviets captured the *Walter* design office together with the *Walter* turbine manufacturing plant and two complete turbines.[19] As was the case with the Western Allies, the U-boats transferred to the Soviet Navy were used to trial the advances in German submarine engineering technology and evaluate if they could be incorporated into their submarine fleets. The number of submarines would be dispersed between the Soviet fleets; however, the majority would be based with the Northern Fleet and were to be a serious threat to Europe's northern flank which had to be met by NATO submarines.

[16] D. H. Russell, 'The Cruise of the "Alliance"', *Journal of Naval Engineering*, Vol. 2, Book 3, (Oct. 1948), 288–291.

[17] E. T. D. Hairs, 'The Arctic Snort Patrol of H. M. S/M. "Ambush"', *Journal of Naval Engineering*, Vol. 2, Book 4, (Jan., 1949), 391–394.

[18] John Jordan, *Soviet Submarines: 1945 to the present* (London, Arms and Armour Press, 1989), p. 22.

[19] Jordan, *Soviet Submarines*, p. 22.

The Soviet Submarine Threat

The worsening political situation between the Western Powers and the Soviet Union led to the creation of the North Atlantic Treaty Organisation (NATO) in April 1949 which allowed for a more effective defence of Europe from Soviet aggression. Throughout the 1950s, figures pertaining to the size of the Soviet submarine fleet varied widely. During a debate in the House of Lords, Viscount Hall mentioned a figure in excess of 360.[20] In a Quarterly Newsletter to senior officers some years later, Mountbatten predicted the Soviet submarine threat by 1960 to be between 599 and 659 submarines.[21] Mountbatten only refers to the new *Whisky*, *Zulu* and *Quebec* conventionally powered (SSK) classes of submarine and disregards 160 older types of submarines; although the Soviets did not build the quantity Mountbatten envisaged, a total of 292 submarines of the *Whisky*, *Zulu* and *Quebec* classes were built during the 1950s. I would suggest that the possible reason the Soviets did not build as many as predicted was due to their limited operational function, changing naval priorities, new submarine concepts and improved weapons systems that were being developed. The Soviets converted some of their *Whisky* class submarines to carry cruise missiles (SSG) which could be used against land targets on the eastern seaboard of the US or against their strike carriers; these complemented the new build *Juliet* class (SSG).[22] Some *Zulu* class submarines were also converted to become the Soviets' first ballistic missile submarines, the missile had a limited range of 150 km. The subsequent *Golf* class (SSB) could carry three R-11FM Scud missiles; all these classes of submarines were conventionally powered and vulnerable because they had to surface to launch their missiles.[23] This was the threat that the Royal Navy's submarine fleet faced from the late 1940s onwards, and nuclear propulsion was required not only to counter conventionally powered Soviet submarines but their predicted nuclear-powered successors.[24]

In August 1958, the Soviets commissioned, *Leninsky Komsomol*, the first of the NATO designated *November* class (SSN).[25] The following year

[20] Parliamentary debate, House of Lords Sitting 11 April 1951, 1950–51; George VI year 15; Columns 229-290, Fifth series Volume 171.

[21] Broadlands Archive, MB1/I300, 14 February 1957, Annex E.

[22] NATO acronym SSG Ship Submersible Guided Missile.

[23] For further information on Soviet submarines, see John Jordan, *Soviet Submarines*.

[24] See discussion on Klaus Fuchs in Chapter II.

[25] Jordan, *Soviet Submarines*, p. 49.

they launched their first nuclear-powered ballistic submarines (SSBN), the *Hotel* class; however, these too could only carry three (R-13) missiles, albeit with an improved range of 600 km. The development and deployment of the *Zulu* (SSB) and *Whisky* (SSG) classes and Soviet developments in the nuclear propulsion field are the possible reference to the "urgency of the international situation" that Rickover noted in his letter to Commander William Anderson USN discussed in Chap. 4. Despite overestimating its strength, the Soviet submarine fleet still consisted of significant numbers; in 1967, its total strength was reported as 392.[26] It can be argued that, until the mid-1960s, the Soviet Navy was anything but a maritime force that acted in support of the "Red Army". However, it is especially true that during the 1960s and 1970s the Royal Navy had to plan its submarine fleet to cope with the strategic and dynamic changes that were likely to happen with the Soviet submarine fleet.

George Hudson, who has written on Soviet naval doctrine, claimed that the building of Soviet submarine fleet was politically driven from the mid-1950s by Khrushchev and the Soviet Army, who were keen to maintain their position as the "defender of the Motherland", naval doctrine was viewed as supporting Soviet land forces and characterised by the notion that the nuclear submarine should form the main arm of Soviet naval power. This dogmatic view, however, was countered by the formidable Admiral Sergey Gorshkov (Commander in Chief of the Soviet Navy 1956–1985), who continuously argued for a balanced fleet to counter the sub-nuclear, rather than the nuclear, threat of war.[27] Indeed, it is reasonable to argue that the limitations imposed on the Soviet Navy, illustrated by its inability to react other than by a nuclear strike during the Cuban Missile Crisis of October 1962, secured Gorshkov's argument which he espoused more forcefully after the fall of Khrushchev. Although the Soviet Navy developed into a "blue water" navy and commissioned two aircraft carriers of the *Kiev* class during the 1970s, it was in its nuclear submarines that the Soviet Union invested its maritime strength with larger and more powerful submarines of all types. From 1961 the Soviet submarine fleet declined slightly in numbers; however, over the next twenty years: '…their

[26] Robert B. Rogers, 'Trends in Soviet Naval Strategy', *Naval War College Review*, Vol. 21, No. 6, (Feb., 1969), 13–29 (p. 27).

[27] George E. Hudson, 'Soviet Naval Doctrine and Soviet Politics, 1953–1975', *World Politics*, Vol. 29, No. 1 (Oct., 1976), 90–113 (p. 103).

collective tonnage trebled to a gigantic one-and-a-half million tons'.[28] By the collapse of the Soviet Union, the submarine strength was calculated by the author and analyst, Norman Polmar, at over 250 submarines (nuclear and conventionally powered).[29] This was the numerically superior adversary that the Royal Navy (and its NATO allies) had to counter throughout the period of the Cold War and it was to do this with the advantages of advanced sonar equipment and methodologies, stealth (through noise-reduction technologies) and improved propulsion systems.

Conversions to "Fast Battery Drive"

America and Britain identified the risk posed by the Soviet use of "Fast Battery Drive" (FBD) based on the designs of Type XXI U-boat and had, by 1947, begun to respond to the threat. Although already developing nuclear propulsion, in the interim, the Americans began their "GUPPY" programme to convert some of their World War II-era submarines.[30] Eighteen *Tench* class and thirty-four *Balao* class submarines were selected to be streamlined and given increased battery capacity; some submarines were also selected to be fitted with schnorkel masts. As well as achieving greater endurance, the GUPPY conversion programme typically increased the dived speed of these submarines from 8.75 knots to between 16 and 18 knots depending on the type of conversion and the class of submarine.[31] Britain, too, began to look at ways of improving its fleet of World War II-era submarines.

As noted earlier, by 1947 the Admiralty had come to the conclusion that rather than rely solely on radar and improving the tactics of its surface fleet, the best way to counter and destroy the fast submarine was with another submarine and the concept of the hunter-killer was born. A preliminary study into converting some of the World War II 'T' class or the new build 'A' class submarines was organised by Vice Admiral M. Mansfield, Flag Officer Submarines (FOS/M). The study found that converting the

[28] Bryan Ranft & Geoffrey Till, *The Sea in Soviet Strategy* (Basingstoke, Macmillan, 1989 2nd Edition), p. 119.

[29] Norman Polmar, *The Naval Institute Guide to the Soviet Navy, Fifth Edition* (Shrewsbury, Airlife Publishing Ltd, 1991), Table 21.1, p. 92.

[30] US Acronym GUPPY: Greater Underwater Propulsive Power.

[31] For full details of the US Navy's GUPPY programme see John D. Alden, *The Fleet Submarine in the U. S. Navy: A Design and Construction History* (London, Arms and Armour Press, 1979).

older 'T' class submarines would be more practicable than converting the new 'A' class due to the former already having three battery sections and a fourth could easily be fitted under the control room by utilising the oil fuel tanks sited there. One of the naval constructors, A. J. Sims, argued that a better option would be to insert an additional section in the pressure hull aft of the control room as this would alleviate the crowding experienced in the 'T' class control room and make room for additional motors to be fitted.[32] The idea was taken forward and the Director of Naval Construction produced two conversion schemes for the Admiralty to select from.

In Scheme A, much the same work as was conducted on HMS *Seraph* was to be done, external torpedo tubes, fittings and the gun were to be removed, the bridge redesigned to incorporate the revised layout of periscopes and masts, including fitting the snort induction mast, and having bow shutters fitted etc. An additional battery section was to be fitted in lieu of the oil fuel tanks under the control room. These improvements would increase the dived speed from 9 knots to 14.5 knots and the conversion work was estimated to cost £330,000.[33] Scheme B was more ambitious, apart from the recommendations of Scheme A, the proposals covered in Scheme B included inserting an extra length of pressure hull aft of the control room bulkhead, fitting an extra pair of motors and other improvements to the propulsion system. This scheme gave a dived speed of 16.75 knots and increased endurance over the rival scheme, the cost of this conversion was estimated at £410,000.

The Admiralty opted for the latter scheme and plans were put in hand to convert eight of the all welded 'T' class submarines between November 1948 and June 1956. Three of the first four submarines, *Taciturn*, *Thermopylae* and *Totem*, had a fourteen-foot section inserted; HMS *Turpin*, for reasons not discovered, had a twelve-foot section inserted.[34] In 1958, it was disclosed in the Commons that HMS *Turpin* had cost £420,000 to construct during the war and that the price of her conversion had been £1,240,000, well in excess of the estimated cost detailed above.[35] As a result of serious concerns about the sea-keeping capabilities of the

[32] Kemp, *The T-Class Submarine*, p. 131.
[33] Kemp, *The T-Class Submarine*, p. 132.
[34] Critchley, *British Warships Since 1945: Part 2*, p. 34.
[35] Parliamentary debate, House of Commons sitting 14 May 1958, 1957–58; Elizabeth II year 7; Columns 21–36, Fifth series Volume 588.

first conversions, a number of experiments were conducted and it was concluded that sea-keeping would be greatly improved by lengthening the section to be inserted to seventeen feet and six inches (Critchley quotes twenty feet) and this was approved in the case of the last four conversions, *Tiptoe*, *Truncheon*, *Tabard* and *Trump*.[36] The benefits of streamlining, identified in exercises conducted between HMS *Truncheon* and HMS *Alcide*, resulted in Admiralty approval for the limited streamlining for some of the hulls of the older riveted submarines, effectively Scheme A.[37] Five of these submarines were converted—*Tireless*, *Tapir*, *Talent*, *Teredo* and *Token*—the higher capacity battery giving a twenty-three percent increase in power which related to an increase of thirty to forty miles underwater endurance throughout the speed range.[38] The conversions must be deemed a success as most were not scrapped until the early 1970s, some even continued in service with the Israeli navy. As Kemp noted: '…the T Conversions were the Royal Navy's first operational fast submarines and experience gained in them was invaluable in operating the *Porpoise/Oberon* boats as well as the nuclear-powered hunter-killers'.[39]

HTP and Air-Independent Engines

Although the 'T' class conversion programme gave these submarines increased speed and a greater dived endurance, they still required to either snort or to surface to recharge their batteries, and the Admiralty was still keen to continue investigating other areas in their quest for air-independent engines. Along with a number of Type XXIs which the Admiralty had commissioned for trials and evaluation purposes, the Admiralty also possessed a Type XVIIB submarine, *U1407*. This submarine had been scuttled at Cuxhaven in northern Germany in early May 1945, she was salvaged and returned to Kiel prior to being allocated to Britain, she was towed to Barrow in August 1945.[40] In May 1945, the inventor of the *Walter* tur-

[36] Kemp, *The T-Class Submarine*, p. 135.

[37] TNA, ADM 1/25252, Submarine versus submarine attacks: report of trials, 1952–53, Report, 21 April 1951.

[38] Kemp, *The T-Class Submarine*, p. 137.

[39] Kemp, *The T-Class Submarine*, p. 142.

[40] For further details on the Type XVIIB *U1407*/HMS *Meteorite* see National Archives ADM 173/21005-ADM 173/21817, also Derek Waller, The U-Boats that Surrendered. U-1407 (HMS Meteorite) in the Royal Navy – 1945 to 1949. https://uboat.net/articles/97.html [Accessed 27 October 2017].

bine, Dr Helmut Walter, and his senior staff had also been captured at Kiel. By November 1945, the *Walterwerke* factory at Kiel had been stripped of its spares, drawings and other assets which were sent to Britain to support the repair and trials of *U1407*. Dr Walter and several of his engineers were brought to Britain in January 1946 to continue work, under British contracts, at Barrow. A major refit was conducted on *U1407*, the torpedo tubes were removed, all electrical equipment was refurbished, and the ventilation system overhauled. The submarine commissioned 26 August 1947 as HMS *Meteorite* and trials began in 1948 (Fig. 2.1).

The Admiralty required a trials report to evaluate whether it would be in the interest of the Royal Navy to further develop the HTP turbine. The report highlighted the operational prospects of the *Walter* turbine and noted the huge advantages of speed. However, disadvantages noted were the cost of the HTP fuel, about £150 per ton; the requirement for a dedicated HTP supply ship due to the fuel being extremely hazardous; and a depot ship for accommodation, etc.[41] Throughout her trials HMS

Fig. 2.1 HMS/m *Excalibur* being launched 25/03/1955

[41] R. M. Bailes, "'H.T.P.' for Submarine Propulsion', *Journal of Naval Engineering*, Vol. 7, Book 3, (Jul., 1954), 301–03 (p. 301).

Meteorite was fitted with only one turbine, not two as designed, so could only achieve fourteen knots which, while it impressed the British, was not its designed top speed of twenty-five knots.[42] The trials however, gave the Admiralty the confidence to order two new British designed HTP experimental submarines, HMS *Explorer* and HMS *Excalibur*. Concerns were raised in the House of Lords over the cost of HTP fuel; Viscount Hall advised the House that the Americans had successfully launched their first nuclear submarine and the results confirm their expectations of nuclear propulsion. Hall added that, unless the same advantages were to be gained from experimenting with HTP, '…we ought to think twice before we go on with this new kind of propulsion unless fuel is available to a much greater extent and can be produced at a much lower price'.[43]

During 1952, HTP fuel prices were falling, Friedman quotes the cost of a submarine patrol using HTP at about £15,000 for 300 tons of fuel. A comparison for a conventionally powered diesel-electric submarine, which required battery replacement after about six patrols, was around £120,000, equal to £20,000 per patrol. [44] At that time it can be seen that this was a good financial reason for the Admiralty continuing with the trials of HTP-powered submarines. HMS *Explorer* was the first submarine to be a joint managed project between Vickers-Armstrongs (Engineering) at Barrow and the Fleet Engineers Department. The Admiralty Development Establishment, Barrow (ADEB) was formed to facilitate this joint venture and, as Vice Admiral Horlick noted, the experience gained was later invaluable in designing and building HMS *Dreadnought*.[45] HMS *Explorer* was launched 5 March 1954, and during the next two years was alongside at Barrow being fitted out and readied for the first of class sea-trials planned for April 1956. As proposed in the trials of HMS *Meteorite*, the crew were accommodated in a dedicated tender, an old mine-layer *Miner VIII*. They also had a dedicated HTP fuel supply ship, RFA *Sparbeck*. HTP refers to hydrogen peroxide in concentrations greater than eighty percent which can be very stable with some materials, such as stainless steel and plastics. However, it is a highly combustive material and can be very explosive if it

[42] Derek Waller, https://uboat.net/articles/97.html

[43] Parliamentary debate, House of Lords sitting, 27 July 1954, 1953–54; Elizabeth II year 3; Columns 121–190, Fifth series Volume 189, Col. 132.

[44] Norman Friedman, *The Postwar Naval Revolution* (London, Conway Maritime Press, 1986), p. 196.

[45] Horlick, 'Submarine Propulsion in the Royal Navy', p. 4.

comes into contact with most metals, hydro-carbons, clothes and skin.[46] Indeed, in June 1955, the submarine HMS *Sidon* sank alongside the depot ship HMS *Maidstone* at Portland when a torpedo, powered by HTP, exploded killing a number of the crew.[47] The same fate befell the Russian submarine *Kursk* in August 2000. As such, HTP requires special precautions when handling and a ready supply of fresh water to dilute any spillages.

HMS *Explorer* was designed so that the HTP fuel was carried external to the pressure hull in four groups, each consisting of four bags on either side of the hull. The tank, pipework and pump on board *Sparbeck* were manufactured from stainless steel and the procedure for fuelling was to pump the fuel into a header tank on board the *Sparbeck* and allow the HTP to gravity drain into the bags. On first filling during the summer of 1956 the HTP began to decompose which resulted in two explosions requiring a docking period for repairs.[48] In early spring 1957, HMS *Explorer* eventually conducted trials in the Irish Sea, however, problems were experienced with carbon monoxide and carbon dioxide leaking into the turbine room, these problems were never fully resolved. Small fires also tended to break out in the turbine room so the compartment was unmanned and the machinery controlled remotely during operation, these problems earned the boats the nicknames *Exploder* and *Excruciator*. The speed log fitted to HMS *Explorer* was modified from an 'A' class submarine and was not deemed to be accurate above twenty knots; therefore, during her full power trials it was arranged that the submarine would keep a steady course and launch a signal at set intervals. The escort frigate would sight these and be able to make accurate predictions of her speed.[49] During trials, the estimated top speed was calculated to be near thirty knots, faster than the USS *Nautilus*, but only available whilst the 100 tons of HTP fuel lasted. Evidently, lower speeds could be maintained for longer, and dived endurance would increase as less fuel is required; it was stressed, somewhat disingenuously, in the House of Lords that her high speed would be sustainable.[50]

[46] A. S. C. Sanderson, 'German Hydrogen Peroxide U-Boat Engine', *Journal of Naval Engineering*, Vol. 1, Book 1, (Apr., 1947), 84–86 (p. 86).

[47] Critchley, *British Warships Since 1945: Part 2*, p. 32.

[48] Brian Hall, 'HMS *Explorer/Excalibur*, 1956–62', *The Naval Review*, Vol. 94, No. 2, (May, 2006), 132–42 (p. 135).

[49] Hall, 'HMS *Explorer/Excalibur*, 1956–62', p. 139.

[50] Parliamentary debate, House of Lords sitting, 27 July 1954, Col. 181.

However, before the trials were conducted in HMS *Explorer*, and her sister, HMS *Excalibur*, the Admiralty was supporting research into nuclear-powered submarines. The two submarines continued to be utilised in training the Royal Navy's surface and submarine fleets to attack fast submarines. The experience gained in manufacturing material of the cleanliness standard required for use with HTP, the welding techniques developed for stainless steel and other disciplines and skills learnt with the two submarines were of great benefit to Vickers-Armstrongs and the Royal Navy. The staff with their newly acquired skills transferred from ADEB to the Dreadnought project when Vickers-Armstrongs was awarded the contract to build HMS *Dreadnought*. In 1948, Captain (E) L. A. Taylor, on the staff of FOS/M wrote in reference to future policy concerning the snorting submarine and the use of HTP: 'Both these are considered to be "stepping stones" to the time when atomic power is a practical proposition for submarines'.[51] Indeed, by 1948, research at Harwell was already being conducted into nuclear propulsion.

The Discovery of Nuclear Fission and Its Application to Submarines

Knowledge of harnessing and exploiting the power of the atom really began in 1938 when the German Nobel winning chemist (1944) Otto Hahn and his assistant Fritz Strassman discovered that they could produce an isotope of barium by bombarding uranium atoms with neutrons, Hahn described the process as "bursting". However, Hahn could not fully explain the results of the experiment conducted 17 December having expected to artificially produce a heavier element. Exclusively informed of the results by Hahn, it was his Jewish assistant, a physicist at the time exiled in Sweden, Lise Meitner, who with her nephew and colleague, Otto Frisch, first fully appreciated the phenomenon of fission and its relation to Einstein's formula, $E = mc^2$.[52] Published in the journal *Nature* in February 1939 (in which Frisch coined the term fission), their theory quickly spread. The theory offered that by absorbing a neutron the uranium atom split into approximately two equal lighter elements (isotopes of elements ranging from selenium—atomic number 34 to lanthanum—atomic number

[51] Taylor, 'Submarine Engineering Problems and Development 1939–1945', p. 133.
[52] Rowland Pocock, *Nuclear Ship Propulsion* (London, Ian Allen, 1970), p. 10.

57; uranium's atomic number being 92). The theory also proposed that the fission process released further neutrons with a great release of energy.[53]

In the late 1930s, the *College de France* was a centre of excellence for nuclear research. Frederic Joliet-Curie, also a Nobel winning chemist (1935), and his colleagues demonstrated that fission of uranium was indeed accompanied by the release of further neutrons and estimated it to be between two and four per fission.[54] Joliot-Curie and his team noted that moderating the speed of the neutron increased the probability of fission; however, they were unable to promote a chain reaction. Only after further research and experimentation was it realised that it was not natural uranium, ^{238}U, which released its neutrons. It was discovered that it was the rare isotope ^{235}U, which constitutes about 0.7 percent of natural uranium, which releases its neutrons upon being bombarded during the fission process. By May 1939, the conditions had been established for maintaining a chain reaction and patents had been filed in Paris, for a proposal for a nuclear reactor. By the end of the year when France was at war with Germany, Joliet-Curie and his team were instructed by the French Minister of Supply, Raoul Dantry: '…to continue their work with the object of developing a submarine engine which did not need oxygen'.[55] In this instance, Joliet-Curie was the first scientist to lead a team to tackle the problem of submarine nuclear propulsion.

At the same time of these great discoveries in Europe, in the US, during discussions of Hahn's discovery at a Conference of Theoretical Physics held in Washington D.C. in January 1939, Niels Bohr and Enrico Fermi (Nobel physicists 1922 and 1938, respectively) proposed that if uranium atoms underwent the process of fission then the energy released would be enormous and there was a possibility that sufficient neutrons could be released to form a chain reaction. During the spring of 1939, Fermi attended a meeting with Admiral S. C. Hooper USN, technical assistant to the US Naval Chief of Naval Operations, Fleet Admiral William D. Leahy USN, to explain the potential of these recent discoveries. Fermi, however, was cautious and reluctant to predict the possibility of a chain reaction without further data. One man without such qualms was the head

[53] H. D. Smyth, 'Fifty Years of Atomic Physics', *Proceedings of the American Philosophical Society*, Vol. 90, No. 1, (Jan., 1946), p. 5.

[54] Valery Nesvizhevsky, Jacques Villain, 'The discovery of the neutron and its consequences (1930–1940)', *Comptes rendus de l'Academie des sciences throughout history*, Vol. 18, Issues 9–10, (Nov–Dec., 2017), 592–600 (p. 597).

[55] Pocock, *Nuclear Ship Propulsion*, p. 12.

physicist at the US Navy's Naval Research Laboratory, Ross Gunn, who fully appreciated the possible application of a controlled nuclear reaction for submarine propulsion.[56] Gunn requested a meeting with one of the US Navy's principal naval engineers, Chief of the US Navy's Bureau of Engineering with responsibility for the Naval Research Laboratory, Vice Admiral Harold G. Bowen USN. It was Bowen who wrote the first detailed treatise on the application of nuclear power to submarine propulsion in November 1939, '*Memorandum on Sub-Atomic Power Sources for Submarine Propulsion*'.[57] Holmquist and Greenbaum suggest that Gunn requested and received $2000 of funding from Bowen.[58] However, Hewlett and Duncan, and other authors, note that the best Bowen could offer was $1500, which Gunn allocated to his friend, Merle Tuve, at the Carnegie Institution in Washington for further investigation of the fission process.[59] In another article, it is suggested that the Carnegie Institution agreed to carry out the work but declined the funds because of internal policy reasons.[60] What is clear is that in early 1939, the US Navy was interested to learn more about nuclear fission and its possible use as a power source for submarines.

In the UK, Sir George Thomson, Nobel physicist (1937), had read the articles published by the above scientists. Thomson requested and received from the Air Ministry the loan of a ton of uranium oxide for his own experiments at Imperial College.[61] Thomson and other British scientists too were aware of the possibilities of the military application of nuclear fission but had established by the outbreak of war that a chain reaction '…was not possible using uranium oxide and ordinary water or paraffin as the second constituent'.[62] When war was declared, Frisch happened to be in the UK visiting Birmingham University to meet Rudolph Peierls, a

[56] Carl O. Holmquist and Russell S. Greenbaum, 'The Development of Nuclear Propulsion in the Navy', *Proceeding of the United States Naval Institute*, Vol. 86/9/691, (Sept., 1960), 65–71 (p. 66).

[57] Pocock, *Nuclear Ship Propulsion*, p. 11.

[58] Holmquist and Greenbaum, 'The Development of Nuclear Propulsion in the Navy', p. 66.

[59] Hewlett and Duncan, *Nuclear Navy*, p. 17.

[60] Lawrence Badash, Elizabeth Hodes, Adolph Tiddens, 'Nuclear Fission: Reaction to the Discovery in 1939', *Proceedings of the American Philosophical Society*, Vol. 30, No. 2, (Jun., 1986), 196–231, p. 213.

[61] George Thomson, 'Anglo-U S. Cooperation on Atomic Energy', *American Scientist*, Vol. 41, No. 1 (Jan., 1953), 75–80 (p. 76).

[62] Thomson, 'Anglo-U S. Cooperation', p. 76.

German physicist who had elected not to return to Germany when the Nazis came to power in 1933. Both men were surprised to discover how little enriched uranium was required to produce an atomic bomb and, in March 1940, they wrote the *Frisch-Peierls Memorandum* which was delivered to the British government to forewarn them of the possibilities.[63] At the Air Ministry and later the Ministry of Aircraft Production, the MAUD Committee was formed to discuss and report on the possibilities of an atomic bomb, the *Frisch-Peierls Memorandum* was discussed and commented upon in their 1941 report.[64] Although some British scientists were now aware of the possibilities of an atomic bomb and of harnessing the slow release of fissionable energy, at that time Britain did not have the economic or industrial resources of America, and its priority during 1940–1941 was one of survival.

During World War II the majority of funds for nuclear research in the US were allocated to the Manhattan Project for the development of the atomic bomb. Hewlett and Duncan noted: 'While the Army had been spending $2.5 billion in building a nation-wide complex of nuclear laboratories, production plants and reactors, the Navy was permitted to do little more than preliminary development of a secondary process used to produce fissionable material for the atomic bomb'.[65] It is evident that given the allocation of resources and priorities during World War II, the US Navy's research into nuclear reactor engineering was severely curtailed. At the end of World War II, apart from the nuclear research facility at Chalk River, Canada, where most of the British scientists and engineers carried out their research, the nuclear industry in the UK was practically non-existent. Due to the successful security policy of Brigadier General Leslie Groves, the head of the Manhattan Project, of excluding direct British scientific involvement, Britain was ignorant of the means of large-scale production of enriched uranium and plutonium. Furthermore, the effect of the US McMahon Act in 1946 would have an immediate impact on British nuclear ambitions, preventing the transfer of nuclear information from the US to other parties. In isolation from its former partner, Britain would have to overcome and solve the chemical, metallurgical,

[63] Dalitz, R. H., 'Peierls, Sir Rudolph Ernst', *Oxford Dictionary of National Biography (1907–1995)*, Oxford University Press, 2004; online edn, https://doi.org/10.1093/ref:odnb/60076

[64] Department of Scientific and Industrial Research: Report for the Year 1947–8, With a Review of the Years 1938–48; 1949 (Cmd. 7761), The Atomic Bomb 3.2.1.

[65] Hewlett and Duncan, *Nuclear Navy*, p. 15.

engineering and other technological problems in order to produce its own enriched uranium and plutonium.[66]

Discussions on the Development of the Nuclear "Engine"

At the end of World War II, the general public became aware of the destructive use of nuclear power with the dropping of an atomic bomb on Hiroshima on 6 August 1945 and three days later on the city of Nagasaki. They were, however, largely ignorant of its potential application for peaceful purposes. Shortly after the War, articles were written promoting the peaceful use of nuclear energy for power stations, industry and for medical research but these articles would mainly have been read by academics and engineers with a particular interest in the subject matter.[67] As a means of introducing the subject to the public, the BBC broadcast a number of interviews with people ranging in disciplines from Professor Mark Oliphant FRS (member of the RNSS and Tube Alloys Project) and Bertrand Russell OM FRS (philosopher and early CND campaigner) to Group Captain Leonard Cheshire VC (an official UK witness to the bombing of Nagasaki and founder of the Cheshire Homes). A book was later published which contained these, and additional thoughts, on the challenges of nuclear energy.[68] Lectures were also given; Sir John Cockcroft spoke of: 'The application of nuclear energy to mobile power units [...] Ship propulsion would seem to offer a more favourable field'.[69] In *Atomic Challenge*, Professor P. M. S. Blackett FRS also postulated the use of nuclear propulsion: '...for very large ships, such as our great liners'.[70] Indeed, in 1947, on their Third Programme, the BBC broadcasted a dedicated talk on the

[66] Alfred Goldberg, 'The Atomic Origins of the British Nuclear Deterrent', *International Affairs (Royal Institute of International Affairs 1944-)*, Vol. 40, No. 3, (Jul., 1964), 409–29 (p. 415).

[67] For example George Thomson, 'Nuclear Energy and its Uses', *International Affairs (Royal Institute of International Affairs 1944-)*, Vol. 22, No. 3, (Jul., 1946) 315–325 and J. R. Menke, 'Nuclear Fission as a Source of Power', *Econometrica*, Vol. 15, No. 4, (Oct., 1947) 314–334.

[68] *Atomic Challenge: A Symposium* (London, Winchester Publications Limited, 1947).

[69] J. D. Cockcroft, 'The possibilities of nuclear energy for heat and power production', *Proceedings of the Institution of Mechanical Engineers* (1947), 206–11 (p. 211).

[70] P. M. S. Blackett, 'Towards Peace' in, *Atomic Challenge*, p. 94.

subject of: '*The Propulsion of Ships by Atomic Energy*'.[71] With all this information entering the public domain there is little doubt that Royal Navy officers and Admiralty scientific staff were also aware of the potential for nuclear propulsion; indeed, some had their own ideas.

In spring 1948, the future application of nuclear power to warship propulsion was discussed after a paper was read by a member of the Royal Corps of Naval Constructors (RCNC), R. J. Daniel, at a meeting of the Institution of Naval Architects in London. Daniel had served with the Royal Navy in the Pacific and had the opportunity to inspect the damage inflicted on Hiroshima and Nagasaki; Daniel was also present as a British observer at the US atomic bomb tests at Bikini Atoll where they had been used against a fleet at anchor in order to assess battle damage.[72] Daniel's lecture was in three parts, in the first section Daniel discussed ship defence against atomic weapons and the third section dealt with the use of a warship armed with nuclear weapons. The middle section of the lecture, however, dealt with the problems of harnessing atomic energy for ship propulsion. Daniel proposed how such a reactor may be constructed and discussed the shielding, metallurgical and other engineering problems that would have to be overcome. Daniel highlighted the advantages to be gained once the difficulties had been resolved. With reference to submarine propulsion Daniel advised that: 'The atomic reactor is well suited to submarine propulsion, developing full power under all conditions, and quite independent of whether the submarine is on the surface or not'.[73] Daniel envisaged that due to the high cost of uranium and the requirement to gain operating experience, it would be several years before nuclear power could be applied to Royal Navy vessels but when the time came it should be: '…devoted to submarine propulsion, for it is in this field that the true worth of this power, its independence of external fuels, is of the greatest value'.[74]

Apart from civilian members of the Institution taking part in the discussions, there were also members of the Royal Navy and the RCNC. It is clear from the discussions that not all members shared Daniel's optimism for nuclear propulsion. D. B. Kimber RCNC argued that until the

[71] 'Broadcasting', *The Times*, 21 June 1947, p. 6, Issue 50793.

[72] R. J. Daniel, 'The Royal Navy and Nuclear Power', *Transactions of the Institution of Naval Architects*, Vol. 90, No. 4, (Oct., 1948), 273–90 (p. 285).

[73] Daniel, 'The Royal Navy and Nuclear Power', p. 280.

[74] Daniel, 'The Royal Navy and Nuclear Power', p. 280.

associated problems of radioactivity had been solved then: '...the possibilities of using this source of power for warship propulsion appear extremely remote'.[75] Vice Admiral Sir John Kingcombe, who, between 1945 until his retirement in 1947, was the Royal Navy's Engineer-in-Chief, was a strong proponent of HTP and is quoted in Waller's article as advising that: '...the whole future submarine policy depends on the successful development of these engines; this is a matter of outstanding importance and urgency'.[76] In the discussions however, Kingcombe called the proposal "Wellsian fiction" and drew the analogy of the promise of unlimited endurance to the amount of potatoes HMS *Nelson*, a World War II battleship, could store. After giving numerous reasons why he thought the proposal implausible, Kingcombe dismisses the idea: 'I suggest that the use of nuclear power in the high-speed submerged submarine will also be impossible'.[77] Kingcombe's dismissal of nuclear propulsion does seem peculiar because in November 1946, the Deputy Controller, John Carroll, and Kingcombe, had stated their intention of seconding as many RNSS personnel as could be spared and two engineer officers to Harwell to form a trained nucleus of a team to take up the development of atomic energy for marine purposes as soon as was practicable.[78] If Kingcombe, as Engineer-in-Chief, held the view that nuclear propulsion was "Wellsian fiction", it may in part explain what appears to have been a half-hearted start to the Royal Navy's nuclear programme. With greater support from the Royal Navy's chief engineer it could be argued that greater resources may have been allocated to the Naval Section and a more forcible argument made for the release of highly enriched uranium to satisfy naval requirements.

Despite Kingcombe's dismissiveness of nuclear propulsion, there were members who spoke in support of Daniel's ideas, although they envisaged the possibility as being anywhere up to fifteen years into the future. Professor A. J. Sims RCNC, noted that many decisions will have to be made before embarking on a project to build a nuclear power propulsion plant, but that: 'The submarine designer, too, has opened up to him at long last the possibility of achieving the true submarine'.[79] Rowland Baker

[75] Daniel, 'The Royal Navy and Nuclear Power', p. 285.
[76] Waller, https://uboat.net/articles/97.html. paragraph 62.
[77] Daniel, 'The Royal Navy and Nuclear Power', p. 287.
[78] DNP 2, NP184/2011, 'The Dreadnought Project Outline Narrative', p. 2.
[79] Daniel, 'The Royal Navy and Nuclear Power', p. 282.

RCNC, who would eventually become the Technical Chief Executive (TCE), head of the successful Dreadnought Project Team at Bath, noted in his written contribution to the discussion: '…that the atomic propelling unit is not in being and cannot be used, at least not for the next few years'.[80] In addressing some of the issues raised, particularly in reference to the time scale involved in building a functioning nuclear propulsion unit, Daniel argued that: 'Surely now is the time that we should begin at least thinking about its application, if not actively planning it'.[81] It is quite understandable that many of the members present at the reading of Daniel's paper may not have been aware of what research and development the Admiralty was undertaking into nuclear propulsion. The following year, the Naval Section at Harwell had investigated the possible use of a helium-cooled reactor and although this did not come to fruition: '…the study provided some useful information about gas cooling for the civil power programme'.[82]

In the years following World War II, it was mainly Royal Navy interest, not political concern, that provided the impetus to improve conventional submarine propulsion. The Admiralty learnt lessons not only from the technological advances of the German Type XXI and XVIIB U-boats, but also in their employment. Twice in the early twentieth century, Britain had nearly lost a war against Germany, its submarine forces inflicting such heavy losses on the British merchant fleet that the country was nearly starved into submission. Although no debates on the subject of nuclear propulsion appeared in Parliament for some time, the subject of submarines was debated, especially when discussing the Navy Estimates. During the late 1940s most of the debates in both Houses of Parliament concerned snorting and HTP submarines. Although there was no political impetus to improve the Royal Navy's submarine fleet, the Prime Minister was briefed on research and development being conducted by the Armed Forces and may have been aware of the Royal Navy's drive towards high-speed submarines and the potential use of nuclear power.[83] Indeed, the first record of a prime minister being informed of the Admiralty's intention to pursue development of nuclear propulsion was not until after 6 April

[80] Daniel, 'The Royal Navy and Nuclear Power', p. 288.

[81] Daniel, 'The Royal Navy and Nuclear Power', p. 289.

[82] Pocock, *Nuclear Ship Propulsion,* p. 127.

[83] TNA, PREM 8/1162, Defence research and production programmes, 1949–50, Report, 10 March 1949, paragraph 12.

1950 when the Admiralty Board had discussed a memorandum, "Nuclear Fuel Submarines"; the First Lord, Viscount Hall, subsequently informed Clement Attlee of this development.[84] It was around this time that the US was about to disclose their intention to build a nuclear-powered submarine and Hall was aware of the possibility of questions being asked in the Lords during an upcoming debate on naval affairs. Hall went to see Attlee to request if he could make a statement to the effect that the Admiralty was working on all new types of submarines.[85] Attlee sanctioned Hall to make his statement in which he highlighted the problem of submerged high speed and endurance and advised that: '...priority has been given to research and development in this particular problem'.[86] From the files researched, however, there appears to be no further correspondence involving a prime minister until Macmillan's request to the First Lord, the Earl of Selkirk, for information on how the nuclear submarine project began, 7 August 1957. This would support the view that nuclear propulsion was not of political interest, which it would become when it was thought that it could be applied to the merchant marine industry.

However, for politicians, and others, interested enough to discover what was happening in the development of submarine propulsion, tucked away in a Command Paper on the Navy Estimates for 1950–1951, it was noted that: 'A programme of investigation into the development of nuclear propulsion is in hand'.[87] This would appear to be the first official public statement that development of nuclear propulsion was being undertaken by the Admiralty. By the early 1950s the subject of nuclear propulsion was seen to emerge and during a debate in the Lords on the Navy Estimates, Viscount Hall made reference to the fact that: '...all possible means of submarine propulsion were under investigation, including systems using nuclear energy and oxygen bearing fuels'.[88] It is to the issue of developing a viable nuclear propulsion plant for submarines that the next chapter is directed.

[84] TNA, ADM 1/26860, Nuclear Fuel Submarines, 1950–53, File cover.

[85] TNA, PREM 8/1244, Proposed development of a nuclear fuel submarine..., 1950, Memorandum, Hall to Attlee, 4 May 1950.

[86] Parliamentary debate, House of Lords Sitting 10 May 1950, 1950; George VI year 14; Columns 225-288, Fifth series Volume 167.

[87] Parliamentary paper, Admiralty statement of the first lord of the Admiralty. Explanatory of the navy estimates; 1950–51 (Cmd. 8160), p. 8.

[88] Parliamentary debate, House of Lords sitting 27 March 1952, 1951–1952; Elizabeth II year 1; Columns 1023–1098, Fifth series Volume 175.

CHAPTER 3

The Nuclear Option

The submarine had nearly defeated Britain in both World Wars; however, the German submarine threat had effectively been resolved in 1943 using radar and attacking the submarine on the surface, or just below, at periscope depth. With the submarine threat foremost in the minds of naval planners at the start of what became the Cold War, the possibility of the nuclear-powered submarine presented a new challenge to naval thinking. Even with a snort mast fitted, the submarine was vulnerable running her diesels just below the surface; during the day, the exhaust and plume from the mast would be visible and improvements in radar technology meant that even an object the size of the snort head was detectable. The nuclear-powered submarine had no requirement to surface so new means had to be found to combat it. Discussions on the concept of the "fighter" submarine had begun in 1947, submarines were not used to protect convoys from enemy submarines in part because friendly forces lacked the equipment to discriminate between friendly and hostile submarines. It could also be suggested, however, that the main problem of using the submarine to hunt and kill other submarines was twofold: the lack of effective information to determine a target's depth and the inability for weapons of that period to be directed in azimuth and, more important, attacking a submarine, in depth onto the target. Indeed, it was acknowledged that air and surface escorts were more effective at protecting a convoy if an enemy

G. M. Jones, *The Development of Nuclear Propulsion in the Royal Navy, 1946–1975*, Security, Conflict and Cooperation in the Contemporary World,
https://doi.org/10.1007/978-3-031-05129-6_3

submarine was discovered within twelve miles of a convoy.[1] The Admiralty paper, TASW 297/49, issued in 1949, determined, however, that the submarine was the best weapon to intercept and destroy enemy submarines but would require a weapon capable of homing in three dimensions or being wire-guided.[2]

The conventionally powered submarine was limited by its speed and endurance; unless lying in wait with a good firing position, the "submersible" would struggle against the "true" submarine. This was made evident during Exercise *Rum Tub* in October 1957, HM Ships *Brocklesby* and *Undaunted*, together with the submarine HMS *Auriga*, were pitted against the USS *Nautilus* in which she beat them convincingly.[3] The rationale in the late 1940s was that a Royal Navy nuclear-powered submarine would be used to train anti-submarine forces against fast-speed Soviet submarines and future nuclear-powered submarines. It would also give the Royal Navy the capability to get the submarine to its patrol area in the Arctic quicker and, thus, spend more time on station to hunt its prey. HMS *Ambush* had been the first Royal Navy submarine to conduct a prolonged patrol in the Arctic using her snort system.[4] There was concern the Soviets would capitalise on the Type XXI U-boat improvements and build them in large numbers. Also of growing concern was the possibility that the Soviets would develop their own nuclear-powered submarines. If the Royal Navy was to fight and win the next submarine war, research and development into nuclear propulsion was essential.

As noted previously, Part II of the MAUD Committee report related to the controlled release of nuclear fission in what they referred to as a "uranium boiler". The committee advised that such a "boiler" could be used to produce power. 'It promises to have considerable possibilities for peace time development, but we do not think that it will be of great value in this war'.[5] Importantly, under the section of "Industrial and other possibilities", the authors noted that using natural uranium with a greater proportion of the isotope ^{235}U would give a chain reaction in ordinary water.[6]

[1] TNA, ADM 1/20414, Report, TASW 297/49, October 1949.

[2] Note: A basic passive homing torpedo, the Mk 20, was introduced into submarines in the early 1950s.

[3] Broadlands Archive, MB1/I300, Quarterly Newsletters, 8 September 1957. See also ADM 1/26756, Operation RUM TUB: analysis, 1958.

[4] Hairs, 'The Arctic Snort Patrol of H. M. S/M. "Ambush"', also Commons Sitting 8 March 1948.

[5] TNA, AB 4/1014, MAUD Committee Report, 15 July 1941.

[6] TNA, AB 4/1014, MAUD Committee Report, 15 July 1941.

The main outcome of the MAUD Committee report was the establishment of the Tube Alloys Directorate within the Department of Scientific and Industrial Research (DSIR), to co-ordinate work such as isotope separation that was required to enrich natural uranium with the isotope ^{235}U to develop an atomic bomb. In the summer of 1942, the project moved to Chalk River in Canada, and in a short space of time became part of the US-led Manhattan Project.[7] From 1946, Admiralty scientists and Royal Naval engineers were appointed to the AERE at Harwell to form the nucleus of a team known as the Naval Section. Their brief was to follow the work of the AERE and to extrapolate information that might be of benefit to marine applications from investigations and experience based on existing reactors. However, prior to 1946 a number of Admiralty scientists were already at work on nuclear matters.

Tube Alloys, US Cooperation and the McMahon Act

The MAUD Committee report, discussed previously, was endorsed by the Scientific Advisory Committee and the War Cabinet accepted the recommendation that work on an atomic bomb should be carried out on a large scale and be given the highest priority. Responsibility for the work was transferred to the Department of Scientific and Industrial Research and, on 1 December 1941, the Lord President of the Council, Sir John Anderson, became responsible for the administration of the project. Wallace Ackers, formerly of Imperial Chemicals Industries Limited, became director of a new department known as Tube Alloys.[8] Work continued at an increased rate at the universities of Birmingham, Cambridge, Liverpool and Oxford; however, it was realised by the end of 1942 that for the work to proceed on the necessary scale it would be unwise to continue in the UK. Therefore, arrangements were made with the Canadian Government for a combined project and during December 1942 and January 1943, some thirty-five British scientists were transferred to continue work in Canada.[9]

At that time, the UK was leading the research field in nuclear fission. In September 1941, Sir George Thomson, chairman of the MAUD Committee, had been given permission to show the MAUD report to

[7] http://www.atomicheritage.org/history/British-atomic-bomb-project [Accessed 4 November 2017].

[8] Cmd. 7761, The Atomic Bomb, 3.2.2.

[9] Cmd. 7761, The Atomic Bomb, 3.2.4.

Vannevar Bush, president of the Carnegie Foundation and head of the US National Defense Research Committee (NDRC).[10] Bush was instrumental in advising Roosevelt of the importance of the British findings and in setting up isotope diffusion plants. Roosevelt agreed that it was: '…desirable to broaden the program, to provide a different organization, to provide funds from a special source, and to effect complete interchange of information with the British'.[11] By mid-1942, the US was spending more and committing greater resources to nuclear fission than the British. To coordinate the atomic bomb effort, the Manhattan Engineer District was formed 13 August 1942 and the following month, 17 September, Brigadier Leslie Groves was appointed to head the project.[12] Baylis has argued that, rather than the complete exchange of information envisaged by Roosevelt, the US military had suggested that the British be given information only where they could make use of it.[13] Indeed, although the Anglo-Canadian project to build a heavy water pile was set up at Chalk River, Ontario, Thomson noted that there was unwillingness amongst the Americans to cooperate.[14]

During 1943, it became evident that the Allies' cause would best be served by pooling resources. Roosevelt and Churchill signed the Quebec Agreement 19 August 1943, through which Britain, and Canada, had entered into a full and effective field of scientific research and development with the US (the Manhattan Project). At the peak of the project, there were fifty British scientists working in the US and one hundred and twenty in Canada, many of whom were RNSS scientific staff.[15] The Quebec Agreement recognised that due to the greater burden falling on the US, any post-war advantages of nuclear energy would be dealt with between the US and Great Britain on terms to be specified by the President to the Prime Minister.[16] Furthermore, under the Hyde Park Memorandum,

[10] Thomson, 'Anglo-U S. Cooperation', p. 79.

[11] Smyth, *Atomic energy for military purposes*, p. 53.

[12] F. G. Gosling, *The Manhattan Project: Making the Atomic Bomb* (United States Department of Energy, Washington D.C., 2010), p. 15.

[13] J. Baylis, *Anglo American defence relations 1939-1980: the special relationship* (London, Macmillan, 1981), p. 17.

[14] Thomson, Anglo-U S. Cooperation', p. 80.

[15] Cmd. 7761, The Atomic Bomb, 3.2.5.

[16] Articles of Agreement governing collaboration between the authorities of the U.S.A. and the U.K. in the matter of Tube Alloys* [*Atom Bomb Research and Development]; 1953–54 (Cmd. 9123).

signed between Roosevelt and Churchill 19 September 1944, it was agreed that full collaboration in developing atomic energy for military and commercial purposes would continue after the defeat of Japan.[17] It is through these terms that the UK envisaged cooperation on atomic matters continuing after the war had been won; however, it was not to be.

Immediately after the war, the US Congress sought to gain civilian, rather than military, control of nuclear energy and to maintain, for as long as possible, a monopoly on nuclear weapons. The US Atomic Energy Act, 1946, was sponsored by Senator Brien McMahon and came into effect 1 January 1947. It has been written that the Hyde Park Memorandum was a private understanding between Roosevelt and Churchill and, being an executive agreement, was not binding on the Truman administration.[18] Indeed, the US copy of the Hyde Park Memorandum had been lost, Roosevelt's naval aide had the copy and a clerk noting the term Tube Alloys, assumed it was something to do with torpedoes, and filed it accordingly. It was not found until 1957.[19] In 1949, the Minister of Defence, A. V. Alexander, had a meeting with Senator McMahon during which McMahon advised that if he had been given all the information which he had requested there would probably have been no need for the Act to have been passed, certainly not one so restrictive.[20] McMahon was eventually shown the British copies of the Quebec Agreement and the Hyde Park Memorandum in 1952. In conversation with Churchill, McMahon explained that the two documents and another document, 'the Declaration of Trust and the Tripartite Agreement of November 1945', signed by Attlee, Truman and Mackenzie King (Prime Minister of Canada) had never been disclosed to Congress.[21]

The Tripartite Agreement came as a result of Attlee writing to Truman on the subject of future international control of atomic energy. The three

[17] Timothy J. Botti, *The Long Wait: The Forging of the Anglo-American Nuclear Alliance 1945–1958* (New York, Greenwood Press, 1987), p. 5.

[18] Barton J. Bernstein, 'The Quest for Security: American Foreign Policy and International Control of Atomic Energy, 1942–1946', *Journal of American History*, Vol. 60, No. 4, (Mar., 1974), 1003–1044, (p. 1022).

[19] Greg Herken, 'London: The Dog That Didn't Bark', in *The Winning Weapon: The Atomic bomb in the Cold War, 1945–1950* (Princeton University Press, 1981). 43–68, (p. 64).

[20] TNA, PREM 8/1097, Minister of Defence forwarded a note about his talk with Senator McMahon on Atomic Energy development in U.S.A., 1949.

[21] TNA, CAB 21/4059, Collaboration with the USA on research and development, 1945–54, Record of Conversation 8 January 1952.

leaders met at Washington 15 November and called for the establishment of a United Nations Commission to control nuclear energy and eliminate nuclear weapons. The following day a memorandum, produced by their advisers, provided: '...that there should be full and effective co-operation among the three nations in the field of atomic energy'; Attlee left Washington the same day feeling he had reached a satisfactory agreement.[22] This final agreement certainly seemed to ensure Britain's continuing cooperation with the US on atomic matters. However, evidence of American resistance to British requests for information mounted during 1946. Matters came to a head in April 1946 when Britain requested information on the construction and operation of atomic energy plants in the US based on the Tripartite Agreement. Truman replied in a forthright manner:

> I must say that no one at any time informed me that the memorandum was proposed with the intention of having the United States obligate itself to furnish the engineering and operation assistance necessary for the construction of another atomic energy plant. Had that been done I would not have signed the memorandum.[23]

The McMahon Act was signed by President Harry S. Truman in August 1946 and came into force the following January. The UK Government's expectation that nuclear information and technology would continue to be transferred under the Agreements discussed above was finally quashed. Truman's dismissal of Britain's atomic ambitions in the immediate post-war period can be seen as a reflection of his attempts during this period to establish a commission in the United Nations to control the production of atomic energy and thus prevent its use for military purposes. It could be argued that Truman's strong support for the North Atlantic Treaty Organisation (NATO), which was formed in April 1949, was an extension of US policy to maintain its monopoly on nuclear weapons by extending the "nuclear umbrella" to protect its Western Europe allies.[24] Certainly, one could reasonably assume that Truman would have thought that the

[22] Goldberg, 'The Atomic Origins of the British Nuclear Deterrent', p. 412.

[23] TNA, FO 800/854, Discussions on atomic energy at meeting of Dominion Prime Ministers in London, May 1946. Telegram, T.1601/46, Truman to Attlee, 20 April 1946.

[24] The US's monopoly ended 29 August 1949 with the explosion of the first Soviet Atomic Bomb, Joe 1.

formation of NATO would have curtailed, if not, halted Britain's nuclear research.

The Attlee government faced the realisation that Britain's nuclear research and development would have to continue independently. Professor Margaret Gowing observed that, in 1946, the Government had in principle, given approval: '...for a low-separation gaseous diffusion plant subject to confirmation before more than limited expenditure had been incurred'.[25] The expenditure limit was £500,000, and the plant's function was primarily to economise the use of uranium and help with research into nuclear-powered submarines. The McMahon Act ended any further collaboration, and the low-separation plant's output would not sustain research into nuclear propulsion and development of the atomic bomb. The only immediate military atomic objective in 1946 was the Government's priority, articulated by the Foreign Secretary, Ernest Bevin, for the UK to develop its own atomic bomb with a "bloody Union Jack on top of it".[26] This was the first obstacle posed to the development of nuclear propulsion and it would not be surmounted for nearly a decade with the construction of the high-separation diffusion plant and doubling the output of the low-separation plant allowing a sufficient supply of fissile material for the Royal Navy's nuclear propulsion development programme.

Throughout the period that the McMahon Act was in force, British governments attempted to get the Act repealed and allow exchange of nuclear information and technology, initial negotiations in 1949 eventually failed due to the arrest of the British scientist, Klaus Fuchs, who had been passing on atomic secrets to the Soviets since his time at Los Alamos. Further negotiations were hampered by the US Atomic Energy Commission (USAEC) and the Congressional Joint Committee on Atomic Energy (JCAE). The JCAE was suspicious that transferring information on submarine reactors would enable the UK to also apply the technology to its civilian commercial activities.[27] However, Sir Leonard Owen, reflecting on the first ten years of nuclear engineering in the United Kingdom, wrote of the McMahon Act: 'Looking back, the McMahon Act was probably one of the best things that happened to the technologists of the British Atomic Energy Project as it made us work and think for our-

[25] Gowing, *Independence and Deterrence: Volume 1*, p. 216.

[26] Peter Hennessy, *Cabinets and The Bomb* (Oxford, Oxford University Press, 2007), p. 48.

[27] Botti, *The Long Wait: The Forging of the Anglo-American Nuclear Alliance 1945–1958*, p. 158.

selves along independent lines'.[28] It certainly led to a divergence in civilian reactor design with Britain leading the world with gas-cooled reactors and the US concentrating on water-cooled reactors.

Acquiring an Experienced Team

Admiralty scientists became involved with atomic energy research in the early 1940s, with scientists being removed from Admiralty work and seconded to the top-secret Tube Alloys Project at Chalk River, Canada. Professor Mark Oliphant, a member of the MAUD Committee who led the RADAR team working for the Admiralty at the University of Birmingham, was one of the first to be seconded in early 1943 along with members of his team.[29] It is apparent from the correspondence between the Director of Scientific Research at the Admiralty, C. S. Wright, and the Director of the DSIR, Sir Edward Appleton, that in late 1943 Wright was opposed to releasing further staff from Admiralty research for the Tube Alloys Project. It is entirely probable that Wright had no knowledge of the work being carried out by the Tube Alloys Project and, inevitably, was reluctant to release more of his scientists from what he considered essential war work. Oliphant was pressing for the release of Sayers and Massey, both men were working on the Centimetre Valve Team at Birmingham university.[30] Lord Cherwell, the Paymaster General, and the Chancellor of the Exchequer, Sir John Anderson, had to intervene.[31] Alexander, the First Lord of the Admiralty, wrote to the Chancellor stating it was unreasonable to expect the Admiralty to release these men from their work.[32] It is apparent from correspondence that the men were released; in a reply to Alexander's letter the Chancellor wrote that when these transfers were agreed by the Admiralty: '...it was made clear by us that the transfer of Sayers would be needed at some future date'.[33] At a meeting in February 1944, the Chancellor discussed a request for the release of further members of scientific and technical staffs, for work on the Tube Alloys Project. Alexander indicated that they had already released a number of experts to

[28] Owen, *Journal of British Nuclear Energy*, p. 23.

[29] TNA, ADM 1/14995, Question of transfer of scientists from centimetre valve team to Tube alloys group, 1943, Letter, Wright to Appleton, 20 October 1943.

[30] TNA, ADM 1/14995, Letter, Appleton to Wright, 15 October 1943.

[31] TNA, ADM 1/14995, Letter, Cherwell to Anderson, 26 October 1943.

[32] TNA, ADM 1/14995, Letter, Alexander to Anderson, 28 October 1943.

[33] TNA, ADM 1/14995, Letter, Anderson to Alexander, 11 November 1943.

work on the Tube Alloys Project, including Dr Massey from the Mine Design Section. Massey's proposed replacement, Dr Allibone, and two members of Massey's team, a Mr. Bates and a Mr. Gunn were also being requested for Tube Alloys work.[34] In June 1944, it was agreed that Professor Jack Diamond of the RNSS would be seconded to Chalk River; two other Admiralty personnel were also made available. Two years later, in a general enquiry of Admiralty interest in atomic energy, Vice Admiral McGrigor, Vice Chief of the Naval Staff, was informed that a Mr. Grout and a Mr. Pinniston had also been seconded to nuclear work.[35] It is evident that the Admiralty seconded many of its scientists to Tube Alloys, although some of the requests to transfer RNSS staff were challenged, it is possible that the Admiralty was keeping an eye on the future and getting its staff the experience that would be required, after the war, to investigate the possibilities of nuclear propulsion.

In June 1944, the Chancellor received a note with reference to a future meeting concerning the implications of the Tube Alloys Project for the naval construction programme. The Chancellor was informed that it would not be possible to discuss the implications of Tube Alloys work on the Navy's future construction programme without revealing the present state of the work and: '...also the extent to which our complete freedom of action in the matter is limited both by general political considerations and by the terms of the Quebec Agreement'.[36] As noted earlier, due to the US committing greater resources to the Manhattan Project, the Quebec Agreement recognised that any commercial or industrial post-war advantages involving nuclear energy to the UK would be determined by the US President who could, in effect, limit any information that could be of profitable benefit to the UK. A fortnight later, the Director of the Tube Alloys Project, Wallace A. Ackers, wrote to the Chancellor advising that the Tube Alloys Project was of special interest to the Admiralty because of the: '...possibility that nuclear energy might be used for ship propulsion. This would affect radically the design of naval vessels, especially Battleships, Aircraft carriers and Submarines'.[37] In the same letter Ackers advised that

[34] TNA, ADM 1/17080, Request from Sir John Anderson for release of certain staff by Admiralty Scientific Research Establishments for tube alloy work in America, 1944, Letter, 9 February 1944.

[35] TNA, CAB 126/173, Admiralty interest in nuclear energy for ship propulsion, 1944–46, Letter, Rickett to McGrigor, 4 November 1946.

[36] TNA, CAB 126/173, Note, Chancellor's office, 14 June 1944.

[37] TNA, CAB 126/173, Letter, Ackers to Anderson, 30 June 1944.

the Admiralty was now engaged in research and development for the future construction programme and therefore wanted to take account of any possible revolutionary developments in ship propulsion evolving from the Tube Alloys work. This appears to be the first British document to mention nuclear propulsion and its application to submarines; however, although the Admiralty suggest engaging in work to meet naval requirements at establishments under their control: '...for reasons of Policy, it is not possible, at this time, to undertake work specially directed towards the development of a plant for the production of nuclear power'.[38] Obviously, at that time the Royal Navy was heavily engaged in establishing the Normandy beach-head in France as well as conducting operations in other theatres of war. However, Ackers did propose that discussions should take place with the Assistant Controller for Research and Development and the Director of Scientific Research at the Admiralty on future developments relevant to the use of nuclear energy for power production. Ackers argued that the Admiralty was agreeable to the proposal and it would allow an additional one or two engineers to be seconded to the project.[39]

In January 1945, Ackers wrote to Major Sumner at the Directorate of Atomic Energy in the Ministry of Supply concerning the employment of Admiralty engineers at the Directorate. The Admiralty assumed that their staff were to be seconded, whilst the Directorate considered that they should be transferred. Ackers requested that: 'In view of the arrangement approved by Sir John Anderson in 1944, I would ask you to agree to the secondment plan'.[40] This was an obvious attempt by the Ministry of Supply to poach from the Admiralty highly qualified personnel whose skills were in great demand. The Admiralty was keen to keep their own personnel because as Ackers noted in the same letter, if the development of atomic energy progressed well the Admiralty would want to establish their own team to investigate the naval applications of atomic power and would want their original staff back to lead that research at some future date.

The Government decision to create the Atomic Energy Research Establishment was taken in October 1945 and to base it at a soon-to-be-redundant RAF airfield at Harwell, Oxfordshire. Harwell was chosen as a

[38] TNA, CAB 126/173, Letter, Ackers to Anderson, 30 June 1944.

[39] TNA, CAB 126/173, Letter, Ackers to Anderson, 30 June 1944.

[40] TNA, AB 1/313B, Use of atomic energy for power by the Admiralty, 1945–47, Letter, Ackers to Sumner, 1 January 1945.

site due to its proximity to London, allowing easy access to the Ministry of Supply and the Admiralty. Harwell was also close to Oxford, thus allowing opportunity for the interchange of ideas amongst scientists at the university there.[41] The Establishment's remit was fourfold: first, the provision of scientific and technical information for producing fissile materials; second, investigation into the application of atomic energy for heat and power; third, the production of radioactive materials and separated isotopes for scientific, medical and industrial research; and, finally, the provision of facilities for research into atomic energy generally.[42] The second of these activities would be the Admiralty's focus of attention. Crucially, any Admiralty interests would be subsidiary to the AERE research and would only be considered if they were of benefit to Harwell's own research as was noted in a draft paper on submarine propulsion, discussed at a meeting 13 August 1951. In paragraph 2.2(c) the author noted that the choice of reactor: '...gave promise of yielding information additional to that required for purely naval purposes and therefore could be considered on grounds other than the naval application', and in paragraph 3.5 noted that: 'A water reactor would be solely of interest to the Admiralty'.[43] It is of interest to note that although the Select Committee Report, (132), advised that a research establishment was to be created, there was no Parliamentary debate on its creation. Indeed, as Gowing noted: 'During the whole period of the Labour Government there was not a single House of Commons debate devoted to atomic energy'.[44] Initially, therefore, there was no Parliamentary scrutiny of the Admiralty's nuclear propulsion programme. I would suggest that, even though the Naval Section at Harwell was small in numbers in the period 1946–1953, had a Parliamentary Committee had the chance to examine its members and investigate the likelihood of producing a successful nuclear propulsion reactor, a Committee would probably have advised the Admiralty to defer the project until fissile material was available to make the project more viable.

It has been noted of Harwell that: 'Although it always employed a mix of engineers and scientists, it was in spirit a scientific rather than an

[41] B. M. Adkins, 'The Atomic Energy Research Establishment, Harwell', *Journal of the Royal Naval Scientific Service*, Vol. 3, (1948), 189–94 (p. 189).

[42] Parliamentary paper, Third report from the Select Committee on Estimates; 1946–47 (132), p. 82.

[43] TNA, AB 6/618, Liaison with Admiralty, 1949–51, Draft Paper, Statement on reactors with particular reference to submarine propulsion. (discussed at meeting 13 Aug 1951).

[44] Gowing, *Independence and Deterrence: Volume 1*, p. 51.

engineering establishment, with a strong science base, and with scientists in almost all the key positions'.[45] This is precisely what Rickover had succeeded in avoiding with his management of the US Naval Reactor Programme, as Vice Admiral Sir Robert Hill wrote: 'Right from the start of the naval programme Rickover preached that the aims would be achieved by good engineering rather than by more science…'[46] With a larger budget and greater resources Rickover was able to exploit his engineering and managerial skills to great effect. The Admiralty, however, would have to proceed at a more restrained pace and take advantage of Harwell's scientific advice; moreover, their concepts for submarine propulsion would have to fit in with the AERE's civil research programme and it is here that the British atomic effort begins to diverge from the research route taken by the US.

On 18 February 1946, the Controller of the Navy, Rear Admiral Charles Daniel, submitted a paper to the Sea Lords, "Consideration of Future Naval Development". Under the section on research and development, Daniel wrote:

> 24. All this research and development covers a vast field, and many years may pass before the new navy will emerge. But I believe that it will emerge, and the change from the present to the future, will be as great as the change from sail to steam. For not only have we to consider atomic attack and defence, but also atomic ship propulsion…[47]

This was the first official reference to nuclear propulsion for warships. As noted earlier, in November 1946, the Deputy Controller, Dr Carroll, and the Engineer-in-Chief (E-in-C), Vice Admiral Kingcombe, had expressed their intention to appoint as many RNSS personnel as could be spared, along with two engineers. However, according to the Royal Navy's unreleased record of their nuclear propulsion programme, it was not until early 1948 that two engineers were appointed, and it was recorded that no further RNSS personnel could be spared. Eventually, one of the engineers had to be withdrawn because of staff shortages in the E-in-C's

[45] John Hendry, '*Technological decision making in its organizational context: nuclear power reactor development in Britain* (Cambridge, Management Studies Group, University of Cambridge, 1991), p. 16.

[46] Hill, 'Admiral Hyman G Rickover USN…', p. 3.

[47] DNP 2, NP184/2011, p. 1.

department.[48] This could be construed as evidence of Kingcombe's lack of belief and possible commitment to the nuclear propulsion project. However, evidence from files released by the UKAEA show a steady intake of Royal Navy and Admiralty staff during the period 1946–1948. In January 1946, Diamond became the first Admiralty scientist to be seconded to Harwell and brought with him a lot of knowledge and experience. Sir John Cockcroft took up his appointment as Director of Harwell at the same time; during their time on the Tube Alloys Project they had worked on the joint British/Canadian reactor plant at Chalk River, Canada, where they helped design a moderate-sized reactor consisting of aluminium-canned uranium rods suspended in heavy water surrounded by a graphite moderator (the prototype for the Canadian NRX pile).[49] Diamond was head of the Naval Section at Harwell until March 1953 when he left to take up the post of Chair of Mechanical Engineering at Manchester University.[50]

In a minute to Cockcroft, Diamond advised that the E-in-C was now willing to appoint both naval engineering officers and civilian engineers to Harwell. 'The total Admiralty complement mentioned is between 20 and 30, including Engineer Chemists, Metallurgists and Physicists'.[51] Carroll confirmed to Cockcroft the intention to second as many RNSS staff as rapidly as possible, the E-in-C will send one engineer officer now and reinforce with a second next year, the intention being that they serve two years and their posts will continue to overlap and so provide continuity. Carroll stated the objective was to build an adequately trained team: '…to take up development of atomic energy for naval marine propulsion as soon as this is practicable'.[52] In May 1947, Diamond noted Lieutenant (E) K. B. Clarke was appointed with responsibility for liaising between the Pile Operating Group and the rest of the AERE concerning matters of the experimental facilities in the pile.[53] Also appointed to Harwell in 1947 was Surgeon Commander Wedd who was conducting research into problems associated with the protection of personnel from radiation. Wedd left, in 1948, to set up a school to instruct naval personnel on the implications of

[48] DNP 2, NP184/2011, p. 2.
[49] Pocock, *Nuclear Ship Propulsion*, p. 48.
[50] DNP 2, NP184/2011, p. 2.
[51] TNA, AB 6/81, Minute, Diamond to Cockcroft, 8 November 1946.
[52] TNA, AB 6/81, Letter, R&D.240, 19 November 1946.
[53] TNA, AB 6/81, Minute, Diamond, 19 May 1947.

atomic energy with reference to naval problems.[54] It can be seen from this correspondence that there was a definite commitment from the Royal Navy and Admiralty to build a dedicated team at Harwell. Also evident from the documentation are the recruiting problems faced by the Admiralty due to the general shortage of trained scientists, technicians and engineers. The First Lord, Viscount Hall, noted in his statement on Navy Estimates, March 1947, that progress in building up the numbers of RNSS personnel would, by necessity, come second to the claims of the universities, schools and industry for trained scientific personnel.[55] Industry was the greatest priority for Britain in the immediate aftermath of the war. Britain needed to regain its international markets and export manufactured goods to rebuild its economy. The shortage of trained scientists and technicians was also cited in the Navy Estimates the following February by Viscount Hall as a limitation to the progress that could be made, although Hall did mention that the maximum effort, and highest priority, was being allocated from the resources that are available.[56]

Harwell and the Admiralty

To keep abreast of developments at Harwell, an informal committee known as the "Tea Party" was established by the Admiralty in early 1947 and was formed of members from a number of Admiralty departments.[57] It is interesting to note the use of the term "informal" in the unreleased narrative file, and again it could imply that nuclear propulsion was not a high priority within the Admiralty at that time. In March 1948, however, the Admiralty gave the committee a more formal title, the Atomic Propulsion Working Party (APWP). The Tea Party visited Harwell 12 May 1947; an informal meeting was held between the Party and members of the AERE staff, the visit also included a tour of the Harwell Pile.[58] One of the actions stemming from the meeting was for work being carried out by the E-in-C department on steam generation associated with atomic piles

[54] TNA, FD 1/468, Research workers, 1946–50, Letter, Loutit to Mallenby, 12 August 1948.

[55] Parliamentary paper, Admiralty statement of the first lord of the Admiralty. Explanatory of the navy estimates; 1947–48 (Cmd. 7054), p. 11.

[56] Parliamentary paper, Admiralty statement of the first lord of the Admiralty. Explanatory of the navy estimates; 1948–49 (Cmd. 7337), p. 2.

[57] DNP 2, NP184/2011, p. 3.

[58] TNA, AB 6/81, Minute, Tea Party Visit 12.5.47, 19 May 1947.

to be put on an official basis by means of a request for this work by the Ministry of Supply to the Admiralty. After the Tea Party visit, Captain J. G. C. Given, head of research and development in the E-in-C's department, wrote to Diamond enclosing a draft letter suggesting the form that the official letter should take.[59] Up to this point it would appear that Harwell had no definitive power parameters to work with and could not really proceed without further details from the Admiralty. Obviously, it was for Harwell to design a propulsion reactor but it was the responsibility of the Admiralty to advise the power requirements and design the steam generation machinery. Another action, for Harwell Staff, was to produce a draft assessment within three months on the present position of nuclear propulsion for circulation to the Tea Party for comment. In August, Diamond submitted a paper and in his introduction he describes the construction of the Clinton pile (this was the second atomic pile after the Chicago pile to be constructed) at Oak Ridge National Laboratory, Tennessee, which was cooled by air and the pile at Hanford, Washington, which was water-cooled. Diamond advised that the design of the piles does not permit working at a sufficient temperature to produce power that can be utilised.[60] Diamond's paper focuses on the current work at Harwell and the piles being constructed in the UK. Areas of general concern to the designers such as weight, materials, biological hazards are discussed, as are future developments. One area of particular concern for naval use was the weight of shielding, it was an important question which Diamond notes, as in other aspects of nuclear energy, reliable information must await experience gained: '…much has yet to be learned about shielding and present designs are probably conservative'.[61]

One further detail of the Tea Party visit to Harwell to address was a joint meeting held with the Power Steering Committee (PSC), an internal Harwell body whose members were from different disciplines. It was agreed at the meeting: 'That the fullest possible interchange of information on research and development should take place between Admiralty and A.E.R.E'.[62] It is important to note that the Chairman of the PSC was Klaus Fuchs who was convicted of spying for the Soviet Union in 1950.

[59] TNA, AB 6/747, Boiler design for marine application: collaboration with the Admiralty, 1947–51, Letter, EN/JGC/G/A2, 20 May 1947.

[60] TNA, AB 6/81, Paper, General discussion of the problems of nuclear power for ship propulsion, 25 August 1947.

[61] TNA, AB 6/81, Paper, 25 August 1947.

[62] TNA, AB 6/81, Minute, Diamond, 19 May 1947.

Fuchs was a German communist sympathiser who fled the Nazis in 1933 and continued his education at Edinburgh University. Fuchs was recruited to work on the Tube Alloys Project in 1941 and was eventually posted to Los Alamos in New Mexico returning to the UK in 1946. After Germany invaded Russia in 1941, Fuchs contacted an old communist colleague who put him in touch with a Soviet embassy official.[63] Fuchs began passing information from then on; inevitably, information concerning the early part of the Admiralty's nuclear propulsion research programme would have been passed to his Soviet handlers thereby giving Soviet scientists an insight into Admiralty thinking, and their progress and manner of solving the various technical problems of reactor design. It has been postulated by Timothy Botti and others that Fuchs' arrest led to the collapse of the 1949 negotiations between Britain and America which aimed at improving nuclear collaboration.[64] Although not convicted of spying, another Harwell scientist, Bruno Pontecorvo, went missing with his family in late 1950 eventually reaching Moscow. It is apparent that, together, these incidents would be enough to damage any thoughts for the foreseeable future by the US of atomic collaboration with Britain. Security vetting of personnel in military/politically sensitive posts was not a process yet adopted in this country; it really was a case of the "Old Boy" network in action.

To enable the AERE to formulate their future research and development programme Cockcroft asked Carroll for the Admiralty's views so they could be considered. Carroll replied that nothing had occurred to alter naval interest and the Admiralty's view was that it was crucial to proceed with development so they may know what is achievable and what the advantages and disadvantages might be.[65] Carroll noted that based on the possibilities and the limited knowledge advised in the earlier paper by Diamond, the Admiralty would not express any views on the design of the pile itself but development was certainly worthwhile. It has been noted there was a national shortage of scientific staff and Carroll could not allocate any more scientific staff to the project. Noting this may affect progress Carroll advised that he would see if it was possible to release a physicist to work with the AERE investigating the reduction of shield weights as it

[63] ODNB, Fuchs, (Emil Julius) Klaus (1911–1988). https://doi.org/10.1093/odnb/9780192683120.013.40698 [Accessed 3 February 2018].

[64] Botti, *The Long Wait: The Forging of the Anglo-American Nuclear Alliance 1945–1958*, p. 60.

[65] TNA, AB 6/618, Letter, R&D 22/49, 11 February 1949.

would improve progress with the research.[66] However, six months later it would appear that the Admiralty was still rather non-committal. Diamond wrote to Cockcroft asking why Carroll's letter of Admiralty interest was not now sufficient support. Diamond noted that when written, it was a helpful guide, but the time was approaching when the Ministry of Supply would need to commit finances and effort on a large scale whilst doubts remained surrounding the practicality of nuclear propulsion which: '…can be resolved only by an expression of Admiralty opinion based on real study of the whole problem'.[67]

Diamond suggested that Cockcroft approach Carroll and arrange for the Admiralty to conduct the study with the assistance of the AERE. In October, Cockcroft wrote requesting a meeting with Admiralty representatives to discuss Harwell's reactor programme.[68] The meeting was held at Harwell, 19 November 1949, and discussion centred on whether the advantage of nuclear power to submarine propulsion would be worth the research and development effort. Cockcroft advised that the cost of the first reactor was about £2,000,000 and the fuel was: 'unlikely to be less than £600,000'.[69] As a comparison, £2,500,000 would purchase one of the Royal Navy's new *Daring* class destroyers which were launched between 1949 and 1952.[70] Cockcroft did note, however, that the fuel would still have some monetary value once discharged from the reactor as only ten percent of the fuel will have been burnt, and the "spent" cartridges could be returned to the Ministry of Supply for a considerable sum. Dr J. V. Dunworth, head of Harwell's Nuclear Physics Division, emphasised the need for clear guidance from the Admiralty so time was not lost in asking questions once the project started. It can be seen here that Harwell staff were looking for direction from the Admiralty so they would know roughly what power output the staff should be designing into the proposed reactor. The Admiralty noted there was a greater advantage to be had in a nuclear-propelled submarine than a surface ship, and a tonnage of between 2500 and 3000 tons was mentioned; accordingly, it was proposed that the first prototype should be built with that application in mind. It was also emphasised that any reactor built for sea trials should

[66] TNA, AB 6/618, Letter, R&D 22/49, 11 February 1949.

[67] TNA, AB 6/618, Note, Diamond to Cockcroft, 19 July 1949.

[68] TNA, AB 6/618, Letter, Cockcroft to Carroll, 21 October 1949.

[69] TNA, AB 6/618, Minute, Admiralty visit to AERE, 28 November 1949.

[70] Parliamentary paper, Navy dockyard and production accounts 1952–53 Annual accounts of […]; 1953–54 (119), p. 22.

be of such reliability that only minor problems remained to be surmounted.[71] During discussions on endurance, the Admiralty said that 100 days at full-power equated to two years' wartime service and: '…in view of the expenditure involved the submarine should have a working life of, say, 700 days'.[72] The Admiralty's requirement for its first nuclear-powered submarine to operate for 700 days would give the AERE staff immediate problems concerning the moderator.[73] The proposed reactor would have graphite as the moderator and would be gas-cooled. Although graphite would outlast one or two charges of uranium, its reliability to withstand those operating conditions for 700 days was unknown. Harwell staff advised that an alternative moderator, beryllia, was considered capable of doing so but would add at least £1,000,000 to the cost of the reactor.[74] The "gas-cooled" avenue of research was to be a cul-de-sac development for the Royal Navy, this research was a result of the water-cooled reactor piles at Hanford in Washington State, which had been built to produce plutonium for the Manhattan Project. Analysis had shown these piles to be inherently less safe than similar gas-cooled reactors and, on this basis, given the size of the UK, it was decided to focus British development on gas-cooled reactors.[75]

The Admiralty advised they were studying the problems of propulsion for high-speed submarines, but so far every system proposed was very expensive. The Admiralty had quoted that using HTP at £5000 per hour would cost £12,000,000 if run continuously for 100 days as was being proposed for a nuclear-propelled submarine.[76] This is a rather skewed comparison as an HTP-propelled submarine could only maintain a high speed for a few hours at best due to the amount of fuel that could be carried. A better comparison would be the FBD submarine: 'Battery costs per patrol £20,000 (batteries require replacing after six patrols at a cost of £120,000). HTP costs per patrol are advised at £15,000'.[77] In a paper

[71] TNA, AB 6/618, Minute, Admiralty visit to AERE, 28 November 1949.

[72] TNA, AB 6/618, Minute, Admiralty visit to AERE, 28 November 1949.

[73] Moderator: A substance which slows neutrons in a thermal reactor to enable fission to take place. Gareth Davies, ed., *Glossary of Nuclear Terms, July 2014* www.burges-salmon.com (2014).

[74] TNA, AB 6/618, Minute, Admiralty visit to AERE, 28 November 1949.

[75] Pocock, *Nuclear Ship Propulsion*, p. 50.

[76] TNA, AB 6/618, Minute, Admiralty visit to AERE, 28 November 1949.

[77] TNA, ADM 1/23729, Requirements for an HTP operational submarine, 1952–53, paragraph 10.

released by the Admiralty Chemical Advisory Panel, it advised the cost of fissile material compared with other fuels. The author, E. J. Macnair, noted the cost of fissile material is less important in a submarine since alternative fuels are equally expensive or more so. 'The estimated consumption of ^{235}U or plutonium (at £250/gram) in a 20,000 KW pile is 1g./hr. The corresponding cost of HTP and fuel would be about £900/hr'.[78] Macnair noted the advantages of nuclear propulsion but raised doubts that a pile small enough to operate in a submarine could be achieved much before 1958, adding that research into HTP should continue. It can be surmised from these figures that the high-speed submarine would be expensive no matter what fuel propelled it and this possibly led the Admiralty to change the emphasis from speed to endurance, New staff requirements were issued in January 1953, this requirement advised that if there was no operational reason for high speed, a thermal reactor might be used. Apart from using less fissile material than an intermediate reactor, it had the advantage of having had more work done on it and, therefore, far more was known about this type of reactor and could reduce research costs.[79] At this point it is worth noting that basic hydrophones (underwater sound devices) had been fitted in Royal Navy submarines since World War I. Just prior to World War II, improved "ASDICs" were developed and fitted into Britain's submarine fleet but these were active sonars, emitting a ping to find the target. With the advent in the early 1950s of passive sonars which listened for a target, the submarine's own speed became a determining factor in acquiring the target; the slower the submarine is travelling the greater the chance of gaining the target. With this in mind it is perhaps reasonable that the Admiralty concentrated on speed as the determining factor for the type of reactor it required.

Discussions continued with the problem of shielding. Harwell staff advised that for the proposed reactor, the shielding would weigh around 700 tons, although this could be reduced by 200 tons if the reactor was situated in the bow or stern. The Admiralty dismissed these proposals citing the shafting and steering arrangements aft and the torpedo tubes forward. It was agreed that because shielding was of great importance and both sides could cooperate now, Harwell would prepare a report on the

[78] TNA, ADM 247/79, Note on the possibilities of applying atomic energy to submarine propulsion, 1947, Paper, Group II Fuels and Propellants, October 1947, paragraph 9.

[79] DNP 2, NP184/2011, p.9. Note: See technical definitions for definitions of Thermal and Intermediate reactors.

problem. The Admiralty also stated that if more collaboration were possible: '...the present high security grading within Admiralty would have to be reduced'.[80] No objection was raised, and Carroll was left to organise the appropriate action within the Admiralty. The meeting ended with Harwell advising the Admiralty that the company, Metropolitan-Vickers, had agreed to undertake construction of the reactor if requested; the Admiralty agreed that the firm was suitable. The Naval Section at Harwell had devoted a lot of their attention to the problems of shielding; the calculations for shielding on ship-borne reactors used data from Hanford in Washington State.[81] This use of data from the US illustrates the fact that some nuclear information was available to the design engineers at Harwell through published journals and articles. By August, Price had produced a paper on shielding requirements noting that: '...there is little—apart from the use of the most efficient materials—that can be done to reduce shielding weights appreciably'.[82] In an undated memorandum, Diamond advised that the AERE will shortly start work on shielding experiments for the Admiralty. Diamond noted that this would involve heavy engineering construction and machining and if left to Ministry of Supply resources delays will occur. Diamond advised that to expedite the shielding work: '...it will be in the Admiralty's interest to provide the workshop and possibly some final design office effort'.[83] Shielding studies continued with the development of the reactor and further papers were raised to reflect new thinking and improved calculations.[84]

Initial Considerations

By November 1949, it was already known that a natural uranium-powered reactor would be too large to be employed in Royal Navy submarines; therefore, Cockcroft formed the Enriched Reactor Group to consider using slightly enriched 2:1 uranium in a gas-cooled, graphite-moderated reactor for power production.[85] The initial reactors built at Harwell and Risley were primarily to produce radio-isotopes for research and medical

[80] TNA, AB 6/618, Minute, Admiralty visit to AERE, 28 November 1949.

[81] TNA, AB 6/567, Thermal fission reactor: Admiralty reactor shielding, 1948–52, Note, Price to Diamond, 23 March 1948.

[82] TNA, AB 6/567, Paper, N/M.24, 26 August 1948.

[83] TNA, AB 6/567, Note, Diamond to EJ Hill, circa September 1948.

[84] TNA, AB 6/626, Enriched Reactor Group, 1949–51, Paper, E.R.G.5, 5 June 1950.

[85] TNA, AB 6/618, Minute, The Enriched Uranium Power Reactor, 14 November 1949.

purposes.[86] A minute on the proposed enriched uranium plant noted that: 'Present reactors use natural uranium as fuel and have not been designed so that use can be made of the heat generated in the fission process. [...] They are also much too heavy for naval use'.[87] Royal Naval requirements at Harwell were only investigated if they were to be of benefit to Harwell's primary task which was aimed at gaining information on all types of reactors, noting that land power generation must compete economically with other forms of power generation, such as fossil fuels and hydroelectric. The paper advised that naval requirements demanded fuel with a higher concentration of the isotope ^{235}U but also noted that the same economic caveats do not apply in this case which: '...gives the prospect of naval power being applicable first'.[88] With the formation of the Group, Harwell considered that development of the submarine reactor was a means of obtaining early operating experience with a power reactor.

Dunworth was appointed leader of the Group and Diamond deputy leader, the Group was responsible for initiating and developing the design work.[89] In December, the APWP issued a report of their findings which summarised that a reasonable target for development was to achieve endurance on one charge of fuel, of 100 days at 25 knots in a vessel of 20,000SHP and about 2500 tons.[90] The report noted the precise size of the submarine could not be predicted without building and operating a shore prototype in which the size and weight of the reactor and shielding would be known. However, the report was quite prescient, advising: 'The first reactor designed conservatively on knowledge available prior to its running may be too large for a vessel of less than some 3500 tons'.[91] The report envisaged that, barring unforeseen difficulties, the first submarine might be ready for sea trials in seven to eight years' time. The £2,000,000 figure advised by Cockcroft at the November meeting was recommended as the cost of reactor development, including a suitable heat exchanger. The report dwelt on whether enough enriched uranium would be available to meet the Admiralty's requirements. The AERE stated that it would be able to provide what was proposed for one submarine and starting in 1953 the fuel will be produced in larger quantities than Ministry of Supply

[86] Anon, 'Atomic Energy Research Establishment, Progress at Harwell', p. 263.
[87] TNA, AB 6/618, Minute, The Enriched Uranium Power Reactor, 14 November 1949.
[88] TNA, AB 6/618, Minute, The Enriched Uranium Power Reactor, 14 November 1949.
[89] TNA, AB 6/626, Note, Cockcroft, 25 November 1949.
[90] TNA, AB 6/618, Paper, Nuclear Fuelled Submarines, 19 December 1949.
[91] TNA, AB 6/618, Paper, Nuclear Fuelled Submarines, 19 December 1949.

requirements. It was noted, however, that the extent of supply to meet Admiralty requirements depended upon such factors as the rate of consumption in trials and tests and the production and stockpiling of atomic bombs.

Viscount Portal, Controller of Nuclear Energy, assured the Admiralty that sufficient low enrichment fuel would be available for one submarine to be at sea by 1959 and twelve by 1968.[92] With these assurances the Admiralty had to set out what design and development effort was required to support the project noting this would require a joint effort of the relevant Admiralty departments and staff at the AERE. Portal had been Chief of the Air Staff from October 1940 until December 1945 and it may be argued that, as a former leader of the Royal Air force, he may have had a bias towards meeting their requirements and ignoring naval needs for enriched uranium. However, from the material researched there appears to be no evident partiality in favour of RAF demands, it was a simple case of not enough enriched uranium being produced and government priorities. Portal had reluctantly taken the job when offered by Attlee, but with the attributes he displayed during his wartime career, he effectively managed not only the demands of the Chiefs of Staff, but also the work being conducted under the three giants of UK atomic research, John Cockcroft at Harwell, Christopher Hinton at Risley and William Penney at Aldermaston.

The same APWP report considered the uncertainty surrounding materials, noting that deterioration of the moderator, failure of the fuel sheathing or self-poisoning may also limit the endurance of the reactor. All parts of the reactor system design and engineering proposals would require careful investigation and before an appraisal could be made to determine the suitability of materials to be used. The report noted that to be successful, the design of the reactor, the associated propulsion machinery and the submarine hull must proceed together. Given the very early stage of design and development, the Admiralty appeared to be cautious in committing itself totally to the proposed reactor design noting that during the design and development process other types of reactors may evolve that might be preferable to the proposed type to be developed by Harwell. Following circulation of the APWP report, Carroll wrote to Cockcroft enclosing a proposed statement to forward to the Defence Research Policy Committee (DRPC). Carroll advised that the statement consisted of the salient points

[92] DNP 2, NP184/2011, p. 4.

of the scientific and technical elements of the report, advocating the importance of: '…successful development of such vessels and a firm statement of Admiralty policy'.[93] The statement closed with the recommendation that every effort be made to produce a nuclear submarine. Four days later the Controller, Admiral Sir Michael Denny, and the Assistant Chief of Naval Staff, Rear Admiral Sir Ralph Edwards, wrote to Viscount Hall and the First Sea Lord, Admiral of the Fleet, Lord Fraser, recommending that the Board give its support noting: 'This development offers the Navy a submarine of performance transcending that of any other type…'[94]

Towards the end of February, Cockcroft wrote to Carroll enclosing a copy of a commentary on the Admiralty paper that had been agreed by the Atomic Energy Council. Cockcroft suggested, with minor changes to the paper, that the commentary be submitted with the paper to the Defence Research Policy Committee.[95] The commentary can be seen as a statement of governmental atomic energy policy whose focus was on gas-cooled, graphite-moderated, natural-uranium-fuelled reactors. This was the AERE's area of competence and it made economic and engineering sense to concentrate their effort on reactor designs where there were fewer unknown problems to be tackled. It is with this policy in mind that the Committee had already recommended that: '…the development of a fast or intermediate reactor shall be given first priority in view of its importance to the development breeding processes for nuclear fuel'.[96] The Committee, however, recognised that it might be possible to develop one other thermal reactor in parallel, provided sufficient industrial assistance was available.[97] The Committee advised that development of a high-temperature thermal reactor fuelled with enriched uranium would be an important step in demonstrating the feasibility of nuclear power and its development as a pilot plant for submarine propulsion would influence the form of the development. The Committee therefore advised that: 'The development should only therefore be linked with the naval requirements if the Defence Research Policy Committee evaluation shows that there are worth-while operational advantages to be gained'.[98] It can be seen from

[93] TNA, AB 6/618, Letter, Carroll to Cockcroft, 27 January 1950.

[94] TNA, ADM 1/26860, Nuclear fuel submarines, 1950–53, Letter 31 January 1950.

[95] TNA, AB 6/618, Letter, Cockcroft to Carroll, 24 February 1950.

[96] TNA, AB 6/618, Letter, Cockcroft to Carroll, 24 February 1950. (Enclosed commentary of Paper on Nuclear Fuelled Submarines, AEC/166, 7 February 1950, paragraph 5).

[97] TNA, AB 6/618, AEC/166, 7 February 1950, paragraph 6.

[98] TNA, AB 6/618, AEC/166, 7 February 1950, paragraph 6.

the commentary that Cockcroft and his staff at the AERE was willing to work on a land-based low enrichment fuel reactor as a prototype for a submarine project so long as the Admiralty's requirements suited the AERE's design and development effort.[99] Without doubt, Harwell's determination to concentrate design on intermediate, rather than thermal, reactors and the lack of availability of highly enriched uranium to support research into reactors for naval propulsion would delay the completion of the submarine propulsion project by a number of years. The following year the Ministry of Supply would start work on a High Separation Diffusion plant to increase production of highly enriched uranium. Only when the demand for research and development of the atomic bomb had been satisfied would highly enriched uranium be made available to the Admiralty project.

The Mark I Enriched Reactor

When Diamond met representatives of the E-in-C's department at their offices at Foxhill, Bath, the object was to gain their opinion on the suitability of Metropolitan-Vickers as the choice of contractor for the Mark I reactor. Diamond advised that the AERE was after a list of firms with a realistic capability of undertaking the work and advice on their suitability in order to assess whether another firm appeared superior enough to warrant an enquiry: '...which would possibly jeopardise the arrangements already made with Metropolitan-Vickers'.[100] A number of firms were discussed including English Electric, CA Parsons, British Thomson Houston and Rolls-Royce, and their respective merits and disadvantages debated. Metropolitan-Vickers was listed first in order of preference and advised as "technically very good—administratively difficult". Diamond was advised that the E-in-C had never found the firm easy to deal with: 'Technically they tend to go their own way and ignore advice'.[101] On the merit side the E-in-C advised they had a most competent team and were already working on HTP turbines for submarines. It became evident during the meeting that Metropolitan-Vickers was probably a better choice than the other firms discussed. It is not clear whether staff at the AERE had doubts about the ability of Metropolitan-Vickers to conduct the work or whether,

[99] DNP 2, NP184/2011, p. 4.
[100] TNA, AB 6/618, Minute, Diamond, 24 March 1950.
[101] TNA, AB 6/618, Minute, Diamond, 24 March 1950.

through the E-in-C's advice, they merely wanted the confidence of knowing they were contracting the work to the most suitable firm. What is certain is that queries were being made of all the prospective firms and the AERE was not willing to proceed further without having searched the market. The same day of the meeting, Cockcroft wrote to Sir John Hacking, head of the British Electricity Authority, asking for his advice on the relative merits of the major electrical firms.[102] In April, Cockcroft received a note enclosing a draft paper suggesting he may like to put it to the next Council meeting. The paper advised the Council to reconsider their decision to discuss suitability of firms with representatives of the electrical engineering industry. The reasons given were that discussions on the proposed design study had been ongoing with Metropolitan-Vickers for some months and there was some concern if they found out that enquiries were being made of other firms they may withdraw from the project. Since the previous meeting steps had been taken to get advice from the Admiralty and the British Electricity Authority. 'In each case, quite independently, Metropolitan-Vickers was recommended as the organisation best equipped to tackle the problems involved in this project'.[103] The paper ends requesting the Council to agree to the development contract being placed with Metropolitan-Vickers.

The AERE began holding exploratory meetings with Metropolitan-Vickers from autumn 1949 and further meetings were held in December and February, during which the AERE expanded on their ideas for their reactor programme. After a meeting of Metropolitan-Vickers' chief engineers, Major J. W. Buckley, Metropolitan-Vickers' new product manager, wrote to Diamond to advise that it was not felt possible to do any further detailed design work without more information concerning the core dimensions and operating parameters for the helium.[104] A paper was discussed in June by the AERE laying out the objectives of the Mark I reactor noting, in order of priority, that it was (1) to demonstrate general design and construction techniques for high-temperature reactors, (2) to study the technical feasibility of power generation from gas-cooled thermal reactors and (3) to establish the nuclear characteristics and provide

[102] TNA, AB 6/1269, Feasibility study for the Mk I enriched reactor by Metropolitan-Vickers Electrical Co. Ltd., 1950, Note, Cockcroft to Hacking, 24 March 1950.
[103] TNA, AB 6/1269, Note, Willson to Cockcroft, 6 April 1950.
[104] TNA, AB 6/1269, Letter, Buckley to Diamond, 2 May 1950.

experimental irradiation facilities.[105] The above objectives obviously reflect AERE's civil work priorities, the remaining three objectives note that the intention for Mark I was to be a pilot for Mark II, intended for submarine propulsion, and for it to determine the control features and the best shielding arrangements for Mark II. Originally, the draft specification had Mark I as a pilot for Mark II as its second priority and this can be seen in the copy of the paper from TNA file AB 6/760 (Discussions with the Admiralty); however, in TNA file AB 6/751 (Meetings with Metropolitan Vickers), the priorities have been re-ordered as noted above. This certainly reflects Cockcroft's instructions that the reactor should not be referred to as the naval reactor but the high-temperature or enriched reactor discussed in the next section.[106] It can be seen that the AERE's priorities were not the same as the Admiralty's and it is reflected in the fact that the Admiralty were not willing to pay the full costs for Mark I. The Admiralty advised the Defence Research Policy Committee that although they would not pay for the construction of Mark I they were willing to pay a share of the design costs.[107] Denny hoped that if nuclear propulsion proved likely then the Admiralty would be able to make savings by stopping other current submarine propulsion research and development programmes. The paper discussing the draft specification for the Mark I reactor also defined the areas of responsibility such that the AERE, in consultation with the Admiralty, would specify the operating conditions of the plant. The AERE was to be solely responsible for the physics, site selection, experimental and development work, specification of radiological safety measures and general reactor characteristics and control methods. Metropolitan-Vickers was responsible for the detailed design of the plant, its components and the combined layout of the whole plant together with estimated costs. The Admiralty's responsibility was the design and operating conditions of Mark II and to give similar guidance on Mark I to the AERE and Metropolitan-Vickers so that maximum experience was gained with Mark I for Mark II.[108]

[105] TNA, AB 6/751, The enriched reactor: correspondence and meetings with Metropolitan Vickers Co. Ltd., 1950-52, Paper, Mark I Enriched Reactor, 2 June 1950.

[106] Margaret Gowing, *Independence and Deterrence: Britain and Atomic Energy, 1945–52. Volume 2: Policy Execution* (London, Macmillan Press, 1974), p. 274.

[107] TNA, DEFE 7/2055, Nuclear propulsion for ship and submarine, 1950–58, Minute, DRP(50) 13th Meeting, 13 June 1950, paragraph 9.

[108] TNA, AB 6/760, 'N' Class submarines: discussions with the Admiralty, 1950, Mark I Enriched Reactor Draft Specification for Contract for Stage I(a), 2 June 1950, p. 1.

On 19 June, a meeting was held with Metropolitan-Vickers to discuss what was now referred to as the "Type 'H' Tank" which was to be comprised of the pressure vessel, moderator, cooling channel and fuel cartridges. Diamond informed the meeting that most of the operating conditions for the design criteria for the Main Tank had been settled advising that: 'The maximum heat output is to be taken as being 40 MW for design purposes for all heat exchangers'.[109] At the end of June, Buckley wrote to Diamond to expand on items discussed during the meeting. Buckley stressed the importance of embodying as much of the Mark II conditions in Mark I to prevent wasting design effort and incurring possible delays further into the programme.[110] The project was starting to gather pace; at a meeting in July it was reported that E-in-C considered it important that the machinery for Mark II be shore-tested with Mark I to prove the control arrangements and its performance. This was agreed; one shaft set of Mark II machinery would be produced by the middle of 1955: '...and tested in conjunction with the Mark I reactor'.[111] By the summer it was felt enough had been achieved for a draft contract specification to be raised.[112] The Ministry of Supply also thought that sufficient progress had been made for a press statement to be released notifying the public of the placing of a contract with Metropolitan-Vickers noting that the Admiralty was also associated with the project.[113] Indeed, the AERE invited the press to Harwell on 22 June to view a mock-up of the general machinery layout for a nuclear propulsion unit and to hear of the work progressing at Harwell.[114] During the summer, more meetings were held and correspondence exchanged between the AERE and Metropolitan-Vickers culminating in September with a contract for Stage 1a, which contained the changed AERE priorities discussed earlier.[115]

By November, however, there was concern over the lack of progress with the pressure vessel. Metropolitan-Vickers was invited to a meeting to discuss acceptance of a design study for the Mark II machinery and boiler. The meeting appears to have been acrimonious; in Diamond's opinion the results were nil; Diamond also noted: 'My impression is that the firm are

[109] TNA, AB 6/751, Minute, Discussion of Type H Tank, 19 June 1950.
[110] TNA, AB 6/1269, Letter, Buckley to Diamond, 30 June 1950.
[111] TNA, AB 6/760, Minute, 'N' Class Submarines, 19 July 1950.
[112] TNA, AB 6/751, Draft Specification for Contract for Stage 1(a), 2 June 1950.
[113] TNA, AB 6/1269, Draft press release, 8 June 1950.
[114] 'Atomic Energy Development', *The Times*, 22 June 1950, p. 3, Issue 51725.
[115] TNA, AB 6/1269, Contract, 6/Elect/7605/CB30A, 6 September 1950.

getting anxious about their ability to understand the project…' and this was the cause of their hesitancy in accepting the Admiralty contract.[116] This was possibly caused by the Admiralty bringing too much research work to Metropolitan-Vickers. The following January Dr N. F. Goodway, head of the Extra-Mural Research Division, AERE, asked Diamond if the Admiralty should again be told that: '…with MV's limited staff, pushing machinery design at this stage may well delay the ultimate completion of the programme'.[117] To expedite the pressure vessel design it was decided, with Metropolitan-Vickers' agreement, to ask whether other firms would be interested in collaborating on the pressure vessel. A meeting was held with Babcock & Wilcox in January, involving the AERE and Metropolitan-Vickers; another meeting was held the following day with the firm GA Harvey. Discussions took place with Whessoe Co. Ltd. at the beginning of February to discuss designs for a pressure vessel. Goodway noted in a letter to H. Tongue, Harwell's Chief Engineer, that competitive tender against a fixed design was not possible and each firm had been asked if they would produce their own design studies under contract: '…with a view to producing by mid-June 1951, a design and a quotation for manufacture and supply'.[118] By mid-February, Whessoe had already replied advising that although interested in the project, a lack of staff and facilities meant that they were unable to commit the required resources.[119] At the end of February, Goodway wrote to Buckley noting that Metropolitan-Vickers had advised the AERE they were not willing to quote for the manufacture of the pressure vessel. Goodway advised that the letter was a formal permission to place sub-contracts with Babcock & Wilcox and GA Harvey to produce the pressure vessel design studies.[120] In reply to questions from the Parliamentary Secretary, the E-in-C wrote at the beginning of May that Harwell was finding Metropolitan-Vickers: '…slower than they had hoped.', but that there was no point in reviewing the target dates at this stage, noting that the project was linked to the Mark I reactor.[121]

[116] TNA, AB 6/1269, Minute, Diamond, 8 November 1950.

[117] TNA, AB 6/747, Note, 5657, 24 January 1951.

[118] TNA, AB 6/1270, Feasibility study for the Mk I enriched reactor by Metropolitan-Vickers Electrical Co. Ltd., 1951, Letter, Goodway to Tongue, 13 February 1951.

[119] TNA, AB 6/1270, Letter, MN:Eng:MKM, 16 February 1951.

[120] TNA, AB 6/1270, Letter, Goodway to EMR Division, 27 February 1951.

[121] TNA, ADM 1/22451, Development of machinery for nuclear submarines, 1951–52, Letter, C.P.52601/51, 8 May 1951.

Within a couple of months, the two firms selected as prospective subcontractors for the pressure vessel design study had forwarded their proposals. On 30 April, a meeting was held with Babcock & Wilcox to discuss their proposals for the pressure vessel. It was proposed to manufacture an eighteen-foot diameter vessel from two rings and two ellipsoidal end caps leaving three circumferential welds to be made on site.[122] During the first week in June, GA Harvey's design proposal for a spherical-shaped pressure vessel was discussed. Metropolitan-Vickers expressed the opinion that although there was no real objection to the design, it was considered that Harvey: '...were introducing additional problems in respect to the bending moments incurred by the discharge tubes and gas pipes within the vessel'.[123] The two designs were discussed a week later during which Tongue emphasised the need to keep to designs which were equated to the experience available. The AERE stressed the requirement to avoid the catastrophic loss of coolant by selecting material that reduced the risk of failure. It was also noted that Babcock & Wilcox had experienced only one failure of a class-one fusion-welded vessel, and it was considered enough was known of past failures to prevent future occurrences.[124] Apart from Carroll's paper on nuclear-fuelled submarines, dated 19 December 1949, which mentions the use of materials affecting shielding, there are no further specific files or records, from those researched, to material selection until late 1951. Although there is recognition by all parties that selection of materials will be important to the success of the project, there appears to have been no significant metallurgical research to support the enriched reactor. Instead, there was a reliance on the use of existing material and technology in the hope this would produce quicker results. As recorded earlier, a draft paper on nuclear propulsion was discussed in August 1951 between Diamond, Tongue and Dunworth. They refer to the Mark I reactor design as involving the least extrapolation of present experience which, with existing knowledge, could be guaranteed using enriched uranium.[125] Indeed, references to, and discussions of, metallurgy are only readily apparent from 1955 onwards and this neglect surely impacted on the ultimate rejection of this design of reactor for submarine propulsion.

[122] TNA, AB 6/1270, Minute, 9 May 1951.
[123] TNA, AB 6/1270, Minute, Selman, 12 June 1951.
[124] TNA, AB 6/1270, Minute, Goodway, 18 June 1951.
[125] TNA, AB 6/618, Statement on reactors with particular reference to submarine propulsion, August 1951.

By mid-June, however, the Admiralty felt that sufficient progress had been made to seek Treasury sanction of £500,000 to proceed with the project. In reply to the Admiralty's letter of 19 June, H. J. Orem of the Treasury noted that when the project was considered a year before by the Defence Research Policy Committee, the conclusion was that the design study should go ahead: '...but before proceeding further the Committee should reconsider the scheme in the autumn when an estimate of the additional cost would be available'.[126] The Admiralty's finance department letter was the first indication that the Treasury had received of estimated costs and advised that the DRPC should be allowed to comment prior to Treasury sanction being given. The Admiralty forwarded a paper, D.R.P/P. (51) 50, on nuclear-fuelled submarines to the Committee, which was discussed at their 24 June meeting. The Committee endorsed the proposal to spend up to £500,000 and in light of the Committee's agreement, the Admiralty wrote to the Treasury to sanction the spend.[127] A reply was received a few days later authorising the sum to be spent over a four-year period. Subsequently, the Admiralty placed a contract, C.P.8/52601/51/M.325, with Metropolitan-Vickers. The firm was advised that the design investigation to confirm the feasibility of a nuclear submarine subject to the satisfactory performance of the Mark I reactor (Stage 1) and the development of components to the extent necessary to prove that the nuclear submarine was not unviable (Stage 2) were connected and had to proceed together.[128]

Metropolitan-Vickers and the Problem of Scale

After an initial meeting in December 1949 and a further meeting in February 1950 between the Enriched Reactor Group and Metropolitan-Vickers, development was envisaged in two parts. Stage 1 was to build a land-based reactor of minimum size in which uncertainties could be examined. Stage 2 was to build a prototype submarine propulsion reactor. Although the immediate problem would be the design of Mark I, the AERE was also tasked with forecasting design and performance characteristics for Mark II so the Admiralty could proceed with tactical

[126] TNA, T 225/1022, Development of nuclear submarines, demand for fissile material, 1951–56, Letter, Orem to Binnie, 21 June 1951.

[127] TNA, T 225/1022, Letter, McCarthy to Humphrey-Davies, 7 August 1951.

[128] TNA, ADM 1/22451, Letter, D.O.8/52601/M.325/D.R.P.P., 5 September 1951.

and ship design.[129] At the meeting the following figures were used to illustrate the dimensions envisaged for Mark I, the reactor prototype. Pressure shell inner diameter sixteen feet, length of the shell eighteen feet, internally the core was to be ten feet in diameter and the graphite cylinder outer diameter was to be fifteen feet. The reactor would require to be encased in a shield four feet thick if constructed of water and steel or eight feet thick if constructed of concrete.[130] The Mark I reactor itself would be some 3600–4000 cubic feet in volume and in order to contextualise the extent of the problem, the reactor would have to be reduced in size and weight in order for Mark II to fit into the Admiralty requirement of a submarine hull of 2500 tons displacement. To put the Admiralty's requirement into perspective: the Royal Navy's latest 'A' class diesel submarines built at the end of World War II displaced 1120 tons surfaced and had a 23 feet, 3 inch beam; HMS *Porpoise*, the first of a new class of conventionally powered submarines to be launched in 1956, displaced 1605 tons surfaced with a 26 feet, 6 inch beam.[131] It can be determined from these figures that staff at the AERE and at Metropolitan-Vickers, together with naval architects, constructors and the other Admiralty departments, would have to work very hard to resolve the weight and dimensional issues that confronted them.

The following month the staff from Harwell paid a visit to Fort Blockhouse at Gosport to learn more about the operational requirements of the submarine. Scheduled in the visit was a tour of the submarines HMS *Ambush* and HMS *Aeneas* whose internal dimensions are given as 16 feet in diameter and 272 feet in length.[132] These tours would have further demonstrated to the AERE staff the severe limitations on space in a submarine. At the meeting, machinery pitch limitations were discussed and acceleration parameters suggested; attention was also given to the need to reduce noise to the absolute minimum. The Admiralty stressed that the submarine should be quiet at submerged speeds between two knots and ten knots, although the upper limit had not been fixed. As noted in the previous chapter, submarines of that period would not have met this criterion without major streamlining and noise attenuation mounting of machinery. One major concern with high speed that needed addressing

[129] TNA, AB 6/751, Minute, Diamond, 24 February 1950.
[130] TNA, AB 6/751, Minute, Diamond, 24 February 1950.
[131] Critchley, *British Warships Since 1945: Part 2*, p. 51 & p. 68.
[132] TNA, AB 6/760, Minute, Visit to Fort Blockhouse, 15 March 1950.

was the requirement to stop the submarine quickly if the hydroplanes jammed whilst diving as the submarine would quickly pass its safe diving depth at speed. It was suggested that reactor power should be able to be cut off in ten seconds; however, it was noted that further study was needed to define the reactor shutdown requirement.[133]

In June 1950, the Admiralty submitted a paper to the DRPC in which they stated their support for the development of nuclear-powered submarines and advised the expected tactical advantages such a submarine would have over its contemporaries. The scale of the research and development effort including cost was also advised in the paper with a recommendation that: '…every effort should be made to produce the nuclear submarine as soon as possible'.[134] The submarine's displacement was advised at about 2500 tons yet within a few days, at an informal discussion at Foxhill, E-in-C staff had discussed provisionally lengthening the engine room from 55 feet to 74 feet thus increasing the displacement to between 4200 and 4600 tons. It was agreed that Flag Officer Submarines: '…should be warned that investigation so far indicated that the submarine would be larger, and the performance would fall short of that already quoted'.[135] It is difficult to say from the files researched whether the Admiralty's paper was deliberately misleading the DRPC to get official sanction of the project. What is known is that officers in the E-in-C's department at that time certainly knew of the scale of the problem and the requirement to enlarge the submarine. Initial design investigations had confirmed that the first submarine, known as N.1, would be larger than originally envisaged and have a 25 feet diameter pressure hull, a surface displacement of 3700 tons and a speed of 22 knots.[136] Further studies allowing for factors peculiar to naval service such as shock, pitching and rolling, etc., meant that to maintain criticality the size of the reactor would have to increase so substantially as to make the type of reactor unattractive. This submarine, N.2, would see an increase in the pressure hull to 31 feet and the surface displacement to 4500 tons.[137] The design difficulties encountered by Metropolitan-Vickers and the AERE were threefold: First, the structural properties of graphite under prolonged irradiation was unknown, its shock

[133] TNA, AB 6/760, Minute, Visit to Fort Blockhouse, 15 March 1950.
[134] TNA, DEFE 7/2055, Paper, D.R.P.(50)73, 5 June 1950.
[135] TNA, AB 6/760, Note, 'N' Class Submarines, 9 June 1950.
[136] DNP 2, NP184/2011, p. 5.
[137] DNP 2, NP184/2011, p. 5.

properties were doubtful, its support of the structure under pitching and rolling conditions and thermal expansion properties were also unidentified. Second, the cooling medium to be used was helium and here there were concerns over the high pumping power required, the large diameter of the cooling pipes, helium availability and helium's ability to diffuse through solid material which made it a difficult medium to make leak-tight, a paramount concern in designing a reactor. Finally, the size and complexity of the reactor and the uncertainty of the shielding weight and dimensions were also problematic.

It is recorded in the unreleased narrative that: 'Some work was done on a water moderated and cooled reactor and another submarine design, N.3, was prepared. This had a surface displacement of 2480 tons, a pressure hull diameter of 22ft., and an underwater speed of 22½ knots'.[138] No other references have been found to this work and it was possibly little more than an internal feasibility study by the Royal Navy over concerns of the size of the proposed (gas-cooled) Mark I reactor and inspired by the authorisation, in July 1951, of the construction of a nuclear-powered submarine in the US employing a pressurised water-cooled reactor. Whatever the reason for conducting the work on the water-moderated and water-cooled reactor, it is evident that other types of reactors were being looked at other than those proposed by the AERE and it must have been apparent, from the size of the proposed N.3 submarine, that a smaller reactor could be designed using water. It can be deduced from the above figures of the initial submarine design and the subsequent N.1 and N.2 submarine designs, that the reactor designers were struggling to reduce the size of the reactor, limited in effect by using uranium enriched to only twice its normal value. The Admiralty, conversely, was increasing the size of the projected submarine to accommodate the plant. In contrast to the eighteen feet diameter Mark I reactor: 'The core diameter of the US submarine was thought by the British to be 6 feet'.[139] The scale of the problem was the Admiralty required a reactor three times smaller than it was possible to make with low enriched uranium fuel. A more compact designed reactor was only possible using highly enriched uranium fuel, and at that time in Britain it was a scarce and expensive commodity.

Although the Admiralty had been in the process of securing Treasury funding for the feasibility report, it was apparent to the AERE in July

[138] DNP 2, NP184/2011, p. 7.

[139] Gowing, *Independence and Deterrence: Volume 2*, p. 275 footnotes.

1951 that while progress had been made, the reactor design would not be practicable for an operational submarine. Cockcroft wrote to Sir Henry Tizard, Chairman of the DRPC, to apologise for not being able to attend the next meeting and advised that the AERE was not yet in a position to present their report to the Committee but had made good progress and reached some provisional conclusions which would need to be discussed with the Admiralty. Cockcroft acknowledged that the present reactor design could only be built into a large submarine: '…and it may well be necessary to use a different design, using more highly enriched uranium which we now expect to be available in 1956…'.[140] In September 1951, Diamond wrote a paper in which he acknowledged that the present reactor design would mean the submarine would have to be approximately 5000 tons to accommodate it. Diamond noted that: 'Should fuel of higher enrichment become available, the experience of the present study points to the desirability of liquid coolants to permit component size reduction and to simplify the pumping problem'.[141] By October, Metropolitan-Vickers had been requested to submit their final report, including time and costs, on the feasibility of the Mark I reactor. Some sub-contract work was still ongoing and there were reports by other departments within the AERE to expedite before the final report could be submitted.

Although the enriched reactor initially had a strong naval perspective: 'The technical Committee felt that treatment of the enriched reactor simply as a pilot plant for submarine propulsion would restrict the form of development'.[142] This opinion indicates how the Royal Navy's requirements were initially "taken into account" but when development seemed to be restricted by their requirements Harwell changed the design objectives. Cockcroft instructed that the reactor should no longer be referred to as the naval reactor but as the high-temperature or enriched reactor. The success of the enriched reactor project was dependent upon active Admiralty engagement; together with the AERE they had contracted Metropolitan-Vickers to conduct the feasibility study. However, as Gowing noted, the requirement of Mark II (submarine reactor) had largely been abandoned in favour of the land-based Mark I, which, paradoxically, was

[140] TNA, DEFE 7/2055, Letter, Cockcroft to Tizard, 23 July 1951.

[141] TNA, AB 15/2043, An enriched uranium reactor for submarine propulsion Author(s): J. Diamond J. Smith, 1951.

[142] Gowing, *Independence and Deterrence: Volume 2*, p. 274.

dependent on the requirement for the submarine reactor.[143] This was increasingly unlikely due to the reactor size and weight problems, with the abandonment of the Mark I enriched reactor, from which the Mark II submarine reactor was to evolve. However, the AERE did benefit from the valuable information it had gained on gas technology from the Mark I design programme: 'The entire attention of the A.E.R.E. was now diverted to land-based power production piles and the experience gained from the Mark I reactor was diverted into the design of [...] Calder Hall'.[144] The nuclear submarine project would be removed from a Defence Research Class I project status and Harwell assistance would all but be withdrawn.

In February 1952, Cockcroft had a meeting with the Controller of the Navy, Admiral Denny. An aide memoire written for the meeting noted that: '...whatever reactor is developed for submarine propulsion, it will be of use only for submarines and would not be developed in the same form for any other purpose'.[145] The memoire sets out three types of reactors suitable for development varying from: '...a crude, comparatively certain solution of low efficiency,' (the PWR) to a more highly efficient solution with more uncertainties and a longer time scale.[146] An attached synopsis detailed three options for the Admiralty to deliberate; A—a thermal water-moderated and cooled reactor, B—a thermal water-moderated and sodium-cooled reactor and C—an intermediate, beryllium-moderated and sodium-cooled reactor. It is evident from the memoire that the longer-term solutions of B and C were closer to the AERE's programme and it can be inferred from the memoire that Cockcroft would have emphasised B and C as the AERE's preferred course of action. Denny wrote to members of the Admiralty Board, enclosing a memorandum on the subject of the nuclear-fuelled submarine. Denny advised that the feasibility study had indicated that a graphite-moderated, low enriched uranium reactor would have been too large to be of use to the Royal Navy. Denny noted that a high-pressure water-cooled reactor, as being developed by the US Navy, had the disadvantage of low temperature, and a development programme would be quite different from the remainder of the AERE's reactor programme. Denny further noted that:

[143] Gowing, *Independence and Deterrence: Volume 2*, p. 275.

[144] DNP 2, NP184/2011, p. 9.

[145] TNA, AB 6/1051, Liaison with Admiralty, 1952–56, Aide Memoire, Cockcroft, February 1952.

[146] TNA, AB 6/1051, Aide Memoire, Cockcroft, February 1952.

> The need for a nuclear submarine is not so great, especially in view of our limited capacity to carry out special hull and machinery development as to justify a recommendation to the Ministry of Supply to take extraordinary steps to develop what is considered to be an interim solution to the submarine propulsion problem.[147]

It is apparent that, despite the US Navy proceeding with what the Admiralty considered an "interim solution", members of the Admiralty Board were in no hurry to develop a similar reactor design. This is either an acknowledgement of the Admiralty's limited understanding of reactor design or a realisation that the Admiralty would have to rely on the AERE's expertise and guidance. Notwithstanding the lack of enriched uranium needed to build a prototype pressurised water reactor, eschewing preliminary research and design on this type of reactor at that time certainly delayed the project by a number of years, which could have been avoided.

After discussions with Cockcroft and his staff, Denny noted it had been agreed to change the development plan of the nuclear submarine and it was proposed to concentrate on a liquid-metal-cooled reactor of either the intermediate or thermal type. A memorandum, B.781, was placed before the Board of Admiralty by Denny, explaining that the Ship Design Policy Committee at their meeting on 24 April approved the proposal of the Special Propulsion Sub-Committee: '…that a water moderated reactor, using liquid metal as a coolant and heat exchanger be developed in place of Mark I…'[148] The Admiralty was aware that the Ministry of Supply had authorised the building of a High Separation Diffusion (HSD) plant in 1951 ensuring that highly enriched uranium would be available in sufficient quantity for Admiralty use from 1956 onwards. The AERE and Admiralty also had some information on the USS *Nautilus*: its displacement, size, diving depth and maximum submerged speed. They also had a limited amount of information on *Nautilus*' reactor plant; its power output, operating temperature and pressure, noting that the Submarine Thermal Reactor (STR) had been designed by the Argonne National Laboratory and had been: '…put forward in a feasibility report during the summer of 1948'.[149] All this information had been gleaned from a variety of US published documentation, which included Congressional Committee reports, Atomic Energy Commission reports and press releases.

[147] TNA, ADM 1/26860, Letter, Denny to Cilcennan, 29 January 1952.
[148] DNP 2, NP184/2011, p. 8.
[149] TNA, AB 6/1051, Paper, U.S.A. Nuclear Propelled Submarine, 19 May 1952.

The End of the Beginning

Metropolitan-Vickers' report was submitted in May 1952 and advised that: 'As a result of studies carried out jointly between Admiralty and AERE, it has recently been decided by Admiralty that the reactor would be too large for submarine use…'[150] There are six sections to the Metropolitan-Vickers feasibility report and it was noted that the report would only be of interest to the natural uranium power project. Indeed, the feasibility study had not all been in vain as the experience gained in the Mark I reactor was employed in the design of Calder Hall, Britain's first nuclear power station.[151] In a memorandum to Cockcroft, Diamond noted that there was plenty of money for research and development, £400,000 in the E-in-C's Department alone, but that the rate of development would be slow without clear commitment by the Admiralty in terms of man-power. The AERE felt it was necessary for the Admiralty to loan additional staff to Harwell to expedite the purely naval aspects of reactor development. However, after consultation between the Chief of the RNSS and Denny, the request for staff was rejected: '…the term of the letter indicating that the submarine project was not of sufficient priority for even one or two staff to be transferred'.[152] Diamond further noted that there was considerable doubt among those who would do the work as to whether the effort was worthwhile, the view throughout the Ministry of Supply was that the Admiralty was not serious about the project. Three possible lines of action were outlined: abandon the project, continue with gradual development until the Admiralty defined their choice of reactor with cost estimates and give the project top priority. As far as the AERE was concerned it was for the Admiralty to decide its priorities and advise them accordingly.

Following discussions between the Admiralty and the AERE, Tongue wrote to Cockcroft and advised that the Admiralty would examine the

[150] TNA, AB 15/2186, Metropolitan-Vickers Electrical Co. Ltd.: Enriched reactor Mk I feasibility report, Section I, 1952, Chapter 1.

[151] DNP 1, ALO 02617/97 – 000378 Vol I First Lord's Committee on the application of nuclear power to marine purposes, *The Admiralty News Summary No: 136*, 1 January 1958.

[152] TNA, AB 6/1051, Letter, Diamond to Cockcroft, 20 May 1952.

problems of using sodium as a coolant and would advise if they were satisfied and prepared to accept liquid metal cooling. Tongue also advised that Denny was still adamant that the nuclear-powered submarine: '…was of vital interest to the Admiralty'.[153] A few weeks later, Denny wrote to notify Cockcroft that arrangements had been made for the development of the engineering side of the reactor, and all that entails from it, with the Admiralty Development Establishment at Barrow (ADEB) acting as the main contractors. Denny reiterated that it should be clear in everyone's mind that the nuclear submarine propulsion project: '…is in fact an approved project which in theory has been in hand now for some time'.[154] Cockcroft replied to Denny and advised that the project had been discussed with the Atomic Energy Board and that the Chairman, Lord Cherwell, would like to discuss the programme with Denny; Cockcroft also advised that if the project was to proceed Denny should inform the DRPC and the operational case stated.[155] Despite these reassurances from the Admiralty, for the remainder of 1952 little correspondence appears in either the National Archive files or unreleased files in the offices of the Director of Nuclear Propulsion. This implies that the project was effectively mothballed by both parties pending a serious commitment from the Admiralty who, as we have seen, were not giving the project their highest priority, and an assured supply of highly enriched uranium to support the project. It is fair to assume that the Admiralty could no longer support the design of a reactor that was totally lacking in characteristics required for an operational submarine and, therefore, decided to defer any further involvement until highly enriched uranium was made available for their use.

In February 1953, an Admiralty draft paper to the DRPC was tabled for discussion between the Admiralty and the AERE staff with a view to progressing the project. It was noted more was known about the low-power thermal type of reactor than the high-power intermediate type. The AERE staff stated that while the intermediate reactor was not a scaling up of the thermal reactor: '…if high speed was an important requirement, the thermal development would be a useful step in the ultimate development'.[156] At this point it was envisaged that a high-speed submarine, 20–23 knots, would be propelled by an intermediate-type reactor and a low-speed

[153] TNA, AB 6/1051, Letter, Tongue to Cockcroft, 26 May 1952.
[154] TNA, AB 6/1051, Letter, Denny to Cockcroft, 2 July 1952.
[155] TNA, AB 6/1051, Letter, Cockcroft to Denny, 9 July 1952.
[156] TNA, AB 6/1051, Memorandum, Cockcroft, 24 February 1953.

submarine, 10–12 knots, by a thermal-type reactor. Further discussions noted that parallel research, into the naval thermal reactor, for other purposes would gather data for a naval intermediate reactor. The unreleased narrative file notes that in 1953, the AERE design studies were concerned with two thermal reactors, one water-moderated and water-cooled and the other liquid-metal-cooled. 'No work was being carried out on intermediate reactors'.[157] Before making a final decision to embark on the development of the thermal reactor rather than, or as a step towards, an intermediate reactor, the Admiralty requested the AERE to collaborate in forecasting the trends of cost and manpower for both courses so that the statement could be forwarded to the DRPC for review. The completed paper was forwarded to the DRPC on 25 June by John Lang, Secretary of the Admiralty, setting out the importance of nuclear propulsion to naval power. The paper gave a brief history of the development of nuclear propulsion to date, its advantages and strategic significance to the navy and recommended development of nuclear propulsion to be of national importance. Of the present position, the paper noted that there were three possible reactor types suitable for submarine propulsion and there were sufficient resources to investigate two. The AERE was, in any case, carrying out research for purposes other than naval use.[158]

A provisional programme was outlined and development costs were estimated at between £11.325M for a 10 MW liquid-metal-cooled reactor and £15.125M for a 70 MW pressurised water reactor. The Admiralty proposed that the project be placed in Class II of Defence Projects: '...it should be allocated effort and resources to meet the rate of progress expected, but suffer retardation if more urgent needs demand it'.[159] By reply, the DRPC Committee agreed with the Admiralty's opinions advised in Lang's paper that early work should be placed in Class II and that the Chairman should approach the Paymaster General to issue instructions to the AERE. The Committee also: '...invited the Admiralty to raise the matter again before a development project was begun'.[160] Sir Frederick Brundrett, Chairman of the DRPC Committee, wrote to the Chiefs of Staff to inform them of the DRPC's decision and to request that, if research proceeds to plan and financial resources allow, there would be:

[157] DNP 2, NP184/2011, p. 9.
[158] TNA, DEFE 7/2055, Paper, Sec. Min. 7/53, 25 June 1953, paragraph 5.
[159] TNA, DEFE 7/2055, Paper, Sec. Min. 7/53, 25 June 1953, p. 6.
[160] TNA, DEFE 7/2055, Minute, D.R.P./M(53)8, 2 July 1953, Annex paragraph 6.

'… no objection to the diversion of the requisite fissile material when needed in 1957 or 1958'.[161] In September, Brundrett wrote to Lord Cherwell, to advise on the proposed design studies, one design was the pressurised water reactor and the other a graphite-moderated liquid-metal-cooled reactor.[162] Brundrett seems to have made an error in the latter description, the synopsis of reactor characteristics makes no mention of graphite-moderated reactors; indeed, graphite-moderated reactors had been ruled out, partly because of their susceptibility to underwater shock and partly because: '…it was unlikely that the graphite would have the necessary stability of structure under the intense radiation to which it would be subject'.[163] As noted earlier, the unreleased papers state that Harwell was concentrating on two types of thermal reactors, which clarifies that the second type of reactor to be subject to the design study would be the thermal water-moderated and sodium-cooled reactor, option B on Cockcroft's memoire.[164] Brundrett ended the letter by requesting a formal agreement to the design studies being undertaken at the AERE.[165] Cherwell does not appear to have been impressed by Brundrett's letter, writing to Cockcroft to advise that the proposals required further consideration. 'Perhaps you will have the paper circulated to the Board in due course. I take it there cannot be any immediate urgency'.[166] Cockcroft forwarded a copy of this note to Brundrett on 15 September and advised to leave the matter until Cherwell returned by which time he hoped the Chiefs of Staff would have made their decision on the allocation of fissile material.

The unreleased file notes that now the project was in the lowest priority given to defence projects, hardly any effort could be spared at Harwell for projects not in Class I and little progress could be expected from the reduced Naval Section at Harwell.[167] This approach is apparent in Cherwell's letter to Cockcroft above. With the departure of Diamond, the Naval Section at Harwell consisted of two ADEB engineers, one RNSS metallurgist and a Lieutenant Commander (Engineer), the complement

[161] TNA, DEFE 5/47, Ministry of Defence: Chiefs of Staff Committee: Memoranda. Nos 291-377, Minute, C.O.S.(53)343, 16 July 1953.

[162] TNA, DEFE 7/2055, Letter, 407/023/51, 4 September 1953.

[163] TNA, DEFE 7/2055, Paper, Sec. Min. 7/53, 25 June 1953, p. 6.

[164] DNP 2, NP184/2011, p. 9.

[165] TNA, DEFE 7/2055, Letter, 407/023/51, 4 September 1953.

[166] TNA, DEFE 7/2055, Note, H/I.1821, 8 September 1953.

[167] DNP 2, NP184/2011, p. 9.

reflecting the Admiralty's decision to reduce the research effort pending supplies of highly enriched uranium.[168] Gowing referred to Diamond as: '...the moving spirit in the enriched uranium reactor', a description borne out by the proliferation of papers and correspondence bearing his name.[169] With the decision by the Admiralty, on their part, to cease further research into the enriched reactor it would appear inevitable that Diamond must have come to the conclusion that there was nothing further he could contribute to the nuclear propulsion programme. Although no evidence has been found in the archives concerning Diamond's decision to accept the Chair of Mechanical Engineering at Manchester University, the unreleased files do note that: '...the Admiralty lost the main spring of their effort in A.E.R.E'.[170] As one of the main advocates for the enriched reactor, it would be of interest to know if Diamond was aware that the day before his resignation from the RNSS the US Navy's Submarine Thermal Reactor (STR 1) prototype had gone critical for the first time at Arco in the Idaho Desert on 30 March 1953.[171] The future of submarine nuclear propulsion was shown to be in pressurised water-moderated and water-cooled reactors (Fig. 3.1).

Enriched uranium and submarine propulsion are entwined, especially if using ordinary (light) water as a moderator and coolant. In the early 1950s both enriched uranium and heavy water were scarce materials in the UK; it is recorded that Harwell's chief engineer, Mr H. Tongue, thought that research of a light water reactor would restrict further development because of the limited temperature that could be achieved resulting in limited thermal efficiency.[172] The lack of enriched uranium, coupled with the knowledge that the US was developing two types of thermal reactors (water cooled and liquid metal cooled) makes it understandable that the submarine reactor programme was effectively mothballed until more fissile material became available. It is also evident from the opinions described in this chapter, and of what material was available, that the gas-cooled, graphite-moderated reactor came to dominate the civil power programme and discouraged further research into the submarine (pressurised water) reactor for several years. This is borne out by the observations of Captain

[168] DNP 2, NP184/2011, p. 14.
[169] Gowing, *Independence and Deterrence: Volume 2*, p. 273.
[170] DNP 2, NP184/2011, p. 10.
[171] Hewlett and Duncan, *Nuclear Navy*, p. 184.
[172] Hendry, *Technological decision making in its organizational context*, p. 31.

Fig. 3.1 Submarine Thermal Reactor prototype—STR 1, Arco, Idaho, USA

John Jacobsen (an original member of the Dreadnought Project Team). Some years later in 1958, Jacobsen attended a fortnight's reactor course at Harwell for senior executives and all but one afternoon was devoted to the design and advantages of gas-cooled reactors. Jacobsen recalls that the afternoon on water-cooled reactors was devoted to their disadvantages, and that there was no mention of the intrinsic safety of the negative temperature coefficient. More is laid out in the next chapter on the factors affecting the reactivity of a reactor; however, moderator temperature and density are two factors. As the moderator increases in temperature it becomes less dense, thus slowing down fewer neutrons, resulting in a negative change of reactivity. Therefore, if there is a problem removing heat from the core and the temperature rises, it will produce less power. The negative temperature coefficient is also load following; that is, as the throttles are opened to increase speed, the reactivity of the reactor automatically increases to suit, and vice versa. It is this feature that makes the pressurised water reactor most suitable for submarine propulsion. The problem with the UKAEA during the 1950s can be summed up by Jacobsen's observation of his lecturer that afternoon: 'The man knew little about water cooled reactors and nothing about submarines'.[173]

As we have seen in this chapter, by the end of 1953, the Admiralty's military case in favour of a nuclear-powered submarine had been put to the DRPC for approval and agreement had been given to proceed with a feasibility study on nuclear propulsion with limited funds sanctioned by the Treasury. No political considerations had been given to the programme beyond the involvement of the First Lord and the Parliamentary Secretary. It could be that the government ignored debating the nuclear-powered submarine in Parliament in case it provoked questions on whether research was being conducted into a nuclear bomb; certainly, there had been no debates on either subject in both Houses of Parliament during the period of the Labour Government, 1945–1951. The engineering aspect of the project was proving to be very challenging, solutions to the metallurgical problems identified had yet to be resolved and, indeed, the type of nuclear reactor had still to be identified and agreed. With Harwell concentrating on a type of reactor totally unsuited to service in a submarine, it is no wonder the Admiralty appeared to have been half-heartedly involved, especially with their knowledge that in the early 1950s the US Navy was progressing with a pressurised water reactor. Finally, the lack of highly enriched uranium was an impediment to the Admiralty's nuclear programme. Excluding financial funding which would have to be approved by the Cabinet, engineering-wise, the Admiralty was, in 1953, a long way off producing their nuclear-powered submarine.

[173] Private correspondence, Captain John Jacobsen RN Ret'd, 27 June 2018.

CHAPTER 4

The Pressurised Water Reactor

From mid-1953, the Admiralty's submarine reactor programme was given the lowest priority for defence projects, Class II. Harwell could not afford to spend much effort on the Admiralty programme and with reduced manning of the Naval Section the programme was, in effect, mothballed until highly enriched uranium could be made available. However, this period would also give the Admiralty time to improve their knowledge of reactor systems and shielding technology and to re-evaluate the pressurised water reactor as the plant most suitable for submarine propulsion. There would also be time to forge alliances with industry to harness their engineering expertise and ensure industry was ready to deal with the exacting standards required of nuclear engineering with the introduction of new defence standards and quality control measurements. New materials had to be produced more affordably and means of fabricating components with these materials had to be created. Challenges with the fuel element components also had to be overcome. The Admiralty also had to resolve all the uncertainties of reactor physics relating to calculations concerning criticality which they did by awarding a contract to Rolls-Royce to design and build a zero-energy reactor. It was imperative that the Admiralty programme be promoted to Class I and that a nuclear-powered submarine

G. M. Jones, *The Development of Nuclear Propulsion in the Royal Navy, 1946–1975*, Security, Conflict and Cooperation in the Contemporary World,
https://doi.org/10.1007/978-3-031-05129-6_4

be commissioned in as short a time as possible; the Admiralty was already discussing the possibility of having ballistic missile submarines by 1965.[1]

During the period of the early 1950s, Britain was still experiencing a prolonged period of austerity; goods were being manufactured primarily for the export market and the defence budget was continuously being examined with a view to making cuts in expenditure. The RAF appeared to be the main beneficiaries in defence spending during the early 1950s, a new force of 'V' bombers and other types of aircraft were being produced and air stations built overseas.[2] Army numbers were continuously being reduced (conscription would be phased out from 1957 onwards ending in December 1960); however, it was the Royal Navy that was most under attack from defence cuts as it was argued that if another war broke out in Europe it would quickly turn nuclear and, therefore, there was no role for the Navy in nuclear war. The RAF claimed it could carry out the Navy's role of maritime surveillance from its overseas stations; the Navy was struggling to argue its case and to voice its role in the defence of the nation. Nuclear propulsion, when mastered, would give the Navy a submarine able to counter the Soviet submarine threat as they left their bases in the Barents Sea, and with development of the sea launched ballistic missile, a sea-based nuclear deterrence. The RAF would come to see the nuclear-powered submarine as a threat to the airborne deterrent and would try to undermine political confidence in the project.

The Reactor Technical Challenge

The first requirement for a nuclear reaction is access to fissile material. The fuel mainly used in a reactor is, ^{235}U, an isotope which forms just 0.7 percent of natural uranium, 99.274 percent of which is ^{238}U (the remainder consists of isotope ^{234}U). To make a nuclear reactor small enough to fit into a submarine, natural uranium requires to be "enriched" so that it contains a greater percentage of ^{235}U. With health and environmental issues concerning the resultant radioactivity paramount, to ensure reliability in its operation the reactor must not only be built to the highest standards but use the highest quality materials in its construction. The selection

[1] TNA, ADM 205/102, Admiralty, Office of the First Sea Lord, 1952–54: (8a) Long term plan for the Royal Navy, para. 29.

[2] 'V' bombers refer to the Vulcan, Valiant and Victor aircraft produced during the 1950s to carry the deterrent.

of materials to be used in a reactor, and its component parts, and the quality specification of those materials must be assured and controlled. Quality is absolute and pivotal in ensuring the safe operation of the reactor, as Captain John Jacobsen noted: 'It is one thing to design and engineer an atomic bomb for instantaneous energy release. It is another thing entirely to design and engineer a controllable nuclear reactor and systems to power a submarine for thirty-five years without refuelling'.[3] Before nuclear engineering standards were applied, manufacturers would work to ill-defined work scopes, for example, details in some British Standards were left: '…for arrangement with the manufacturer'.[4] Procedures and processes for quality standards, where they existed, differed widely throughout British industry; the control of these differing standards was also inconsistent, with Admiralty overseers, manufacturers' foremen and shipbuilding production engineers applying various control mechanisms. Uniformity of material specification and the testing of those specifications to give product assurance, together with the control of the raw material, the finished product and its certification would mean a quantum leap forward in production engineering and quality throughout British industry.

Having sourced fissionable material, the next challenge is to transfer the heat away from the reactor core. There are different types of nuclear reactors, such as gas-cooled reactor, integral boiling water reactor and pressurised water reactor, to name a few. Using pressurised water as a coolant and moderator in a reactor was first proposed in a paper by A. M. Weinberg in 1946, and it was chosen by the Americans as the type best suited for application in a submarine.[5] After the false start with the enriched gas-cooled reactor and likely encouraged by the launch of the USS *Nautilus*, the Admiralty began discussing the use of thermal water-cooled reactors in 1953. In 1955, the Naval Section at Harwell confirmed to Cockcroft that the Admiralty would be placing a contract for a pressurised water reactor, using enriched uranium with Rolls-Royce.[6] A basic pressurised water reactor is constructed from the following main components; the reactor pressure vessel which contains the fuel elements, control rods, reflector and

[3] Conversation with Captain Jacobsen, one time MEO of HMS/m *Valiant* and Chairman of the Naval Nuclear Technical Safety Panel, November 2014.

[4] A. H. L. Trapnell, 'Standardization in the drawing office', *Journal of Naval Engineering*, Vol. 15, Book 1, (Jun., 1964), 148–59 (p. 148).

[5] A. M. Weinberg, 'High Pressure Water as a Heat Transfer Medium in Nuclear Power Plants', (Atomic Energy Commission Report Mon P-93, April 1946).

[6] TNA, AB 6/1051, Note, Cockcroft, 28 July 1955.

the moderator/coolant. External to the reactor pressure vessel connected by pipework are the main circulation pumps, pressuriser and heat exchanger, these components constitute what is known as the primary system.[7] Within the heat exchanger the primary coolant heats water which produces steam in what is known as the secondary system. Steam is fed to turbo-generators to produce electricity for the numerous submarine systems reliant on electrical power; steam is also fed to the main turbines which propel the submarine. The steam is then condensed and fed back to the heat exchanger in a continuous self-contained process. The cycle begins with the nuclear chain reaction in the reactor core. As discussed in Chap. 1, when hit by a neutron, each atom of ^{235}U disintegrates to produce a pair of approximately equal isotopes of elements from the middle of the periodic table (ranging from selenium to lanthanum), and additional neutrons and energy. To maximise the chance of fission, the neutron's energy is reduced by use of a moderator and is also directed back into the reactor by the reflector. The choice of fuel element, such as uranium oxide/steel or uranium/zirconium; the choice of moderator, such as water or carbon; and the choice of control medium, such as boron or cadmium, have a direct effect on the design and the criticality of a reactor. The presence of reactor poisons, elements that capture neutrons very effectively, and the reactivity of the neutron flux will also determine how easy a reactor design will be to control. 'These effects and others like the proportion of neutrons which escape the core, contribute to the *reactor problem*, whose solution determines whether a design can sustain a chain reaction without risk of meltdown'.[8]

Material within the reactor pressure vessel should not contain elements (other than the control rods) which are strong neutron absorbers as they will choke the chain reaction. Therefore, component materials must be carefully selected for their suitability for use in a reactor. Apart from the requirement for materials not to adversely interact with each other, their corrosion properties, resistance to heat and irradiation and low neutron capture cross section had to be calculated and subjected to experimentation. How reactor components are manufactured and shaped and the position into which they are placed within the reactor pressure vessel are

[7] For further descriptions of pressurised water reactor components, see Davies, *Glossary of Nuclear Terms.*

[8] Peter M. B. Walker, ed., *Chambers Dictionary of Science and Technology* (Edinburgh, Chambers Harrap Publishers Ltd, 2000), p. 797.

also vital parts of the reactor problem that had to be solved. 'That is in ensuring that just one neutron can survive to continue the chain, no more and no less'.[9]

The early nuclear engineers had the tremendous task of finding the best materials for each application in the reactor plant. They had to define the quality specification, source reliable suppliers and produce a system of examinations to assure that the material supplied was to the design specification. Although the engineering difficulties involved in designing a functional nuclear reactor were appreciated, from what is now known there are instances of what one may call "nuclear naivety". Examples include the use of helium because of its property to diffuse through solid material, the handling of fuel elements and the complexity involved in refuelling. The draft specification drawn up by Harwell staff in 1950 for the Mark I reactor advised: 'The permissible helium leakage from the circuit depends partly on the effort required to dilute it to breathing tolerance…'[10] At a meeting in January 1951, it was recorded that: 'A provisional limit of 1 cu. ft of helium per hour has been placed for helium leakage from the whole circuit…'[11] It is fair to assume that today, a discussion on a "designed" leakage of the coolant for a proposed reactor would not only be unacceptable, the design of the reactor would not even make it to the planning stage. Similarly, there seemed to be *laissez-faire* attitude towards the controlling and handling of fuel elements. In a paper written in 1947, Diamond noted that: 'There is every possibility of storing both used and unused charges [fuel elements] onboard ship'.[12]

One last example is contained in a memorandum written in 1956. Captain R. I. A. Sarell noted that: 'Little difficulty is expected in designing reactors so that they can be recharged by ships staff'.[13] Sarell thought it entirely feasible that ships staff could start the process of discharging the reactor within twenty-four hours of shut down. He added another day to refuel the reactor, all with the assistance of a couple of dockside cranes. It is evident from these examples that, although engineering difficulties had been identified, there could be some degree of risk to health and possible environmental issues if no solutions were found to resolve the envisaged

[9] Walker, *Chambers Dictionary of Science and Technology*, p. 797.

[10] TNA, AB 6/760, Draft Specification for Contract for Stage I(a), 2 June 1950, p. 2.

[11] TNA, AB 6/1270, Notes of discussion at Babcock House, 24 January 1951.

[12] TNA, AB 6/81, General discussion of the problems of nuclear power for ship propulsion, 25 August 1947, p. 3.

[13] DNP 1, ALO 02617/97 – 000378, Memorandum, 035, 14 November 1956.

leakage problems. The difficulties involved with refuelling and the storage of spent fuel rods would also need to be addressed. The reactor plant and systems would be engineered to be as safe as possible and handling procedures would be written to mitigate risk. Quality assurance and quality control would ensure issues such as these were managed and monitored in a structured fashion.

It was not only British industry that would have to change their attitude towards quality assurance and control; the Royal Navy too, would also have to change their approach to quality procedures. The *Journal of Naval Engineering* contains many examples of engineering practice, both good and bad, prevalent in the Royal Navy throughout the 1950s. As an example of not being able to identify and control unmarked material, during her refit in 1955 the submarine depot ship, HMS *Maidstone*, landed thirty-five tons of metals: '…the majority of which could not be identified since it bore no markings'; in the comments section it was advised that: 'The marking of metals for identification had been under consideration for many years'.[14] The problem of identification was obviously known to the naval authorities but the time taken to establish procedures to deal with it highlights to a certain extent the bureaucratic nature of the Admiralty. In another article, on hydraulic contamination, it was concluded that the materials supplied had not been in accordance with the specification.[15] This is another example of the Royal Navy not being able to identify and control material. However, there was also lack of uniformity and consistency in the management of ships' engineering departments. In an article on planned maintenance in HMS *Newcastle*, it was noted that the planned maintenance system had been set up using locally printed forms. The comment section noted that there were five cruisers running their own systems of planned maintenance. Admiralty Fleet Orders were issued to introduce: '…the Admiralty scheme for documentation which includes a system for planned maintenance, together with associated forms and equipment.'[16] The absence of a uniform schedule advising periodicity for planned maintenance of equipment would lead to a rather chaotic approach to the subject. As one correspondent noted: '…the Engineer's Office is usually run

[14] 'Notes from Sea', *Journal of Naval Engineering*, Vol. 8, Book 1, (Jan., 1955), 67–80 (p. 69).

[15] B. D. Manhire, 'Contamination of Gun Mounting Hydraulic Systems', *Journal of Naval Engineering*, Vol. 9, Book 2, (Apr., 1956), 214–19 (p. 219).

[16] 'Notes from Sea', *Journal of Naval Engineering*, Vol. 10, Book, 2, (Apr., 1957), 231–43 (p. 232).

to one particular individual's liking, and rarely, if ever, laid out in a standard manner as any organization should be'.[17] Indeed, in 1953, an earlier report had noted the shortcomings in planned maintenance; identifying that the organisation of maintenance was inadequate, the serviceability of machinery was unsatisfactory and the operational availability of ships was insufficient.[18] However, by the end of the 1950s, Fleet Maintenance Units were set up to assist in organising maintenance periods and planned maintenance schedules were eventually introduced into the Royal Navy for all engineering disciplines, air, marine and weapons. It would be crucial, for the reactor, to have a standard operating procedure and a standardised maintenance schedule for ensuring its optimum performance and safe operation.

Harwell too, had experienced problems with quality control; a pilot circulator manufactured by Metropolitan-Vickers had been received with faults at Harwell. The motor was not properly connected, and the rotor assembly was fouling the stationary parts. In a note addressing these problems the author also advised that the cover studs of the pilot circulator had been positioned according to Metropolitan-Vickers' standard practice and the drawings were accepted by the AERE drawing office. However, the flange designed to take the cover studs had been drilled to Harwell's standard practice and due to the half-pitch difference could not be mounted without remedial work being undertaken.[19] Of the files researched, this was the only recorded incident of failure in Harwell's quality control procedures and is an indication of the level of attention to detail that is required when dealing with product design for nuclear components.

The Brontosaurus in the Museum: Quality Assurance

The Admiralty argued that if Britain was to maintain a Royal Navy, nuclear propulsion had to be adopted and contemporary weapons fitted, if it was still to be a relevant fighting force. If not, it risked becoming obsolete and: 'The purpose of its existence has vanished, and it is merely the skeleton of

[17] C. S. Beard, 'Correspondence', *Journal of Naval Engineering*, Vol.11, Book 3, Dec., 1958), 452–60 (p. 458).

[18] I. J. Lees-Spalding, 'Better Late Than Never', *Journal of Naval Engineering*, Vol.16, Book 1, Dec., 1965), 73–78 (p. 74).

[19] TNA, AB 6/1271, Feasibility study for the Mk I enriched reactor by Metropolitan-Vickers Electrical Co. Ltd, 1952. Note, Bannister, 30 August 1951.

a brontosaurus in a museum'.[20] That risk was a possibility if industry failed to break out of the engineering conservatism that dominated the manufacturing sector in the middle of the twentieth century. In 1940s Britain, few manufacturers would have had the capability of producing the material and equipment that would meet the exacting standards, tolerances and cleanliness required to build a nuclear reactor. The development of fabrication methods and welding techniques for new materials would also have to be learnt. Observation of these higher specifications and standards would have to be audited to ensure compliance; it was noted during World War II that British Standard Specifications were not always adhered to. When they were, and drawings were scrutinised for modern production techniques which also contained details of fit, form and function, production methods improved; as the author noted: 'The days of a "nice easy fit" have gone forever'.[21] Writing some years later Rear Admiral MacLean noted that in 1948: '…our own marine industry was then about ten years behind the times'.[22] That is not to say, however, that everyone else was ten years ahead. Even Rickover had to spend time bullying American industry into accepting the high level of quality assurance that nuclear engineering demanded. Rickover is internationally renowned for his abrasive manner, a quality it could be argued that allowed America to produce the world's first nuclear-powered submarine from nothing in the space of eight years. Rickover would justify his brusque managerial style on the basis that he was so dismayed at the poor engineering standards of American industry that: '…it required an inordinate effort from him to raise these standards to what he considered essential for his nuclear submarines'.[23]

It can be argued that the motor and aircraft industries had every incentive in the twentieth century to improve their products, to have a higher quality and thereby be more efficient, more reliable and safer than previous products, or indeed than their competitors' products. No one would buy a car that is always at the garage for one defect or another and an aeroplane with a bad reputation, for example the de Havilland *Comet*, would soon be out of production. British motor and aeroplane manufacturers either improved the quality and cost of their products or they risked

[20] DNP 1, ALO 02617/97 – 000378, Note, A. S. Oslay, 20 November 1956.

[21] Taylor, 'Submarine Engineering Problems and Development 1939–1945', p. 130.

[22] TNA, ADM 205/106, Office of the First Sea Lord, 1955: (8g) Maritime Strategy: The state of British marine engineering, July 1955.

[23] Alex Smith, *Lock up the swings on Sunday* (Durham, The Memoir Club, 1998), p. 219.

losing business to competitors and, by extension, the loss of the company. By contrast, it can also be argued that the UK marine industries saw no requirement to improve their products; they were still complacent from being one of the world's largest ship producers. In 1912 the world mercantile output was 2,902,000 tons, of which 1,739,000 tons was built in the UK.[24] However, their position as the largest producer had been steadily declining since the end of World War I and in the 1950s that decline was accelerating due to reduced costs and increased reliability in machinery from overseas competitors which possibly prompted MacLean to write his paper. MacLean attributed the maritime engineering industries' weakness to three separate problems. First, the issue of selecting the most suitable machinery designs; second, the problem of obtaining the best designs, which MacLean identified as the consequence of handling as much design work as possible within the industry to the exclusion of the high-quality design potential outside the industry. 'The third problem is the production of a first-class result at an economical price'.[25] Yarrow was the only company with a design office capable of undertaking the major design problems discussed in MacLean's paper. The Yarrow Admiralty Research Department team, Y-ARD, was set up after the war and faced jealousy from the rest of industry. MacLean also noted that other engineering disciplines had been employed to assist in educating the maritime engineering industry without which the Royal Navy would not be able to get the equipment they required at the necessary quality standard. However, MacLean noted that few companies were willing to make the required changes, and few had technical staff with the experience to offer the advice needed. The Admiralty, the UKAEA and Rickover, too, would play their part in developing the marine industry's experience in nuclear engineering and its ability to manufacture products to the quality standards required of them.

The Admiralty did not sit idly on the sidelines waiting for the marine industry to revolutionise its manufacturing and quality processes. Naval Engineering Standards were introduced to industry which they were expected to conform with; these standards imposed quality requirements on a manufacturer and also had documentation to control the material. These standards are reviewed periodically and amended in accordance

[24] Sidney Pollard and Paul Robertson, *The British shipbuilding industry, 1870–1914* (London, Harvard University Press, 1979), p. 249.

[25] TNA, ADM 205/106, (8g) Maritime Strategy, July 1955.

with the latest applicable British and International standards. Material for use in a nuclear reactor plant had separate quality assurance regulations and means of identification and control documentation which is also subject to periodical review. During the early years, within Rolls-Royce, quality documentation was not very formalised, although one could argue that it must have been thought good enough for Rickover to advise the Admiralty to use them as their prime contractor. Different departments had their own documentation and: '…confusion concerning data transfer was possible and did indeed arise'.[26] Lambert noted that it was due to staff professionalism that it did not occur more often, Rolls-Royce eventually produced a standardised quality manual in 1976.

Aside from the Y-ARD collaboration, as discussed in Chap. 1, the Admiralty also collaborated with Vickers-Armstrongs; a combined design and development team known as the Admiralty Development Establishment Barrow (ADEB) was formed, which had facilities for design, development, prototype and production testing. As Horlick noted, ADEB represented a different management method in which a small, highly qualified team was dedicated to a single project: '…from initial design through to production testing'; this was to be the future of all naval projects, but change would come slowly.[27] As things stood in the early 1950s, there were many Admiralty Departments, each with their own demarcation zones (as was the case in industry in general), the AERE and civil engineering companies involved with the nascent nuclear propulsion programme, and all were serving different masters resulting in a rather chaotic management of the programme which would not be resolved until the formation of the Dreadnought Project Team in early 1958 under the management of one man as Technical Chief Executive, Rowland (Roly) Baker RCNC.

Both Y-ARD and ADEB would second men to the Naval Section at Harwell to support the Admiralty and the AERE in the design and development work. In 1953, the Naval Section consisted of a Royal Navy engineer, a RNSS metallurgist and two men from ADEB, the Admiralty decided that more resources needed to be committed to the project if it was to progress. Captain (E) S. Harrison-Smith was appointed to head the section from mid-May 1954.[28] In January 1955, Jack Edwards was appointed as the senior RNSS representative, the first since Diamond had

[26] Lambert, *Rolls-Royce: the Nuclear Power Connection*, p. 60.
[27] Horlick, 'Submarine Propulsion in the Royal Navy', p. 4.
[28] TNA, AB 6/1051, Letter, Mason to Cockcroft, 24 March 1954.

left two years previously.[29] Edwards played a major role in the project and would eventually become professor of nuclear engineering at the Royal Navy College Greenwich. By April, the Admiralty side of the section had increased to eleven personnel: four Royal Navy engineers, five RNSS staff, one RCNC constructor and one man from Y-ARD.[30] As the project began to gather pace, engineers were also seconded to the Naval Section from Rolls-Royce and Vickers-Armstrongs. The UKAEA was also keen to share their experience in nuclear technology with industry. William Strath, Managing Director of the UKAEA wrote to Rear Admiral Rebbeck of Vickers-Armstrongs, to advise that the UKAEA was anxious to disseminate their experience to industry and see them play: '…an increasing part in the development of the country's atomic energy programme'.[31] Vickers-Armstrongs was the Royal Navy's principal company for building submarines and there were other companies in the UK who thought that they too had the technical competence to assist the Admiralty in their nuclear propulsion programme.

Industry Joins the Project

Rolls-Royce's Chairman, Lord Hives, heard that the Admiralty was getting frustrated with the slow response from Vickers-Armstrongs and saw his chance to extend his company's remit to include marine engineering.[32] Certainly, by 1955 the Controller of the Navy, Admiral Ralph Edwards, had been in discussion with both firms.[33] It would appear that industry was hesitant to get involved with the nuclear submarine project; Harrison-Smith wrote to Cockcroft to inform him that the Controller and E-in-C, Vice Admiral Mason, were looking to overcome the difficulties in obtaining assistance from industry, a further meeting was to be held between the Admiralty and the Directors of Vickers-Armstrongs and Rolls-Royce on 11 July.[34] Edwards visited Rolls-Royce accompanied by Lord Weeks, Chairman of Vickers Group, and Major General Dunphie, Chairman of

[29] DNP 2, NP184/2011, p. 10.

[30] TNA, AB 6/1051, Letter, Harrison-Smith to Cockcroft, 16 March 1955.

[31] DNP 6, CP60801, UKAEA collaboration with Admiralty and its contractors in RN nuclear propulsion programme, Letter, Strath to Rebbeck, 5 June 1956. (Enclosed in Letter, Hedger to Barman, 25 October 1960).

[32] Lambert, *Rolls-Royce: the Nuclear Power Connection*, p. 16.

[33] TNA, ADM 205/106, (8g) Note, Edwards to Mountbatten, 21 June 1955.

[34] TNA, AB 6/1051, Letter, Harrison-Smith to Cockcroft, 1 July 1955.

Vickers-Armstrongs. The object of the visit was to consider the best organisation that could be developed to progress the nuclear propulsion programme. The group were shown around by Lord Hives and Edwards recorded that Hives already had some twenty scientists and draughtsmen working on various possible installations.[35] It is of interest to note that, in the same message to the First Lord, discussions had taken place concerning the possibility of putting the prototype engine into the first possible ship available, not necessarily a submarine, for testing. By the end of July, discussions had been so successful that the Admiralty had agreed to place a contract with Rolls-Royce for a pressurised water reactor using enriched uranium. Cockcroft noted that: '…the relationship between Rolls-Royce, the Admiralty and Harwell would need to be agreed'.[36] It is apparent that Cockcroft wanted to delineate areas of responsibility to prevent inefficiency and duplication of effort, but it would take time to get all the agreements necessary in place as it would also involve the firms Vickers-Armstrongs and Foster Wheeler.

In July, Hives had been invited to visit the Westinghouse Company at Pittsburgh, the manufacturers of the US Navy's pressurised water reactors, and was planning a return trip with three Rolls-Royce engineers. Hives reported that the project was to be reorganised with Vickers-Armstrongs as the main contractor and Rolls-Royce and Foster Wheeler as subcontractors. Hives also advised that Westinghouse had agreed to give Rolls-Royce: '…access as commercial friends to their pressurised water reactor work'.[37] At that time it was illegal for US citizens to divulge any information on nuclear propulsion, however, the Admiralty noted that there was no objection to Rolls-Royce advising Westinghouse that they were interested in the proposed visit and would presume that Westinghouse would arrange for all US formalities to be complied with.[38] Hives had informed Cockcroft that if arranged by Admiralty: '…they might get advice and help from Westinghouse'.[39] However, advice and assistance would not be forthcoming as Admiral Strauss, Chairman of the US Atomic Energy Committee, noted: '…there was both legal and political difficulty,

[35] TNA, ADM 205/106, (8g) Note, Controller, Nuclear Propulsion, 20 July 1955.
[36] TNA, AB 6/1051, Note, Cockcroft, 28 July 1955.
[37] TNA, AB 6/1051, Note, Schonland, 13 October 1955.
[38] TNA, ADM 205/106, (8g) Telegram, 192115Z, 20 September 1955.
[39] TNA, AB 6/1051, Note, Schonland, 13 October 1955.

and he had to take account of the position of certain individuals'.[40] This last remark is certainly directed at Admiral Rickover who was "dual-hatted" in the US Navy as head of the Nuclear Power Division in the Bureau of Ships and as the naval liaison officer at the US Atomic Energy Commission's Division of Reactor Development.[41] At that time Rickover was firmly opposed to any nuclear technology exchange with the British as he did not think they were committed enough to developing the technology nor had the engineering discipline in place to make it happen.

Noting that the pressurised water reactor had been selected and the project was being reorganised along the lines advised by Hives, the UKAEA, looking to involve engineering firms with nuclear technology, wrote to the E-in-C, enclosing a draft agreement between the UKAEA and Rolls-Royce and any other company collaborating in connection with the project. The UKAEA considered that the greater part of the reactor and heat exchanger effort would be done by them and that Rolls-Royce had underestimated the amount of work required to be done, noting the work would require greater effort from them than Rolls-Royce have indicated.[42] A few days later Cockcroft noted that the Authority would be asked to provide assistance with core physics, metallurgy of fuel elements, chemical effects of radiation on corrosion of materials, shielding studies and siting facilities for the prototype. It was envisaged that Admiralty and Rolls-Royce staff would be made available to assist in the work and Cockcroft would liaise with Harrison-Smith to discuss requirements.[43] With discussions and draft agreements beginning to formalise the project, the Admiralty wrote to the Treasury in December for approval to place an order with Vickers-Armstrongs for a prototype set of machinery. The Admiralty advised the Treasury that they intended to design a nuclear propulsion plant of 15,000–20,000SHP for use in a submarine and required 40 kg of enriched uranium by mid-1956 for critical experimental work. The Admiralty required a further 80 kg of enriched uranium by late 1958 for the prototype reactor and 40 kg for the submarine by late 1960. The project was foreseen as taking six to eight years at a cost: '...in the region of £10M but I am afraid it may well be quite substantially more'.[44]

[40] TNA, AB 16/1379, Marine propulsion: naval reactor for Admiralty submarine, 1954–57, Note, Cockcroft, 3 November 1955.

[41] Hill, 'Admiral Hyman G Rickover USN', p. 3.

[42] TNA, AB 6/1051, Letter, Perrot to Mason, 7 November 1955.

[43] TNA, AB 16/1379, Note, Cockcroft, 11 November 1955.

[44] TNA, T 225/1022, Letter, McKinnell to Radice, 8 December 1955.

Individual access agreements were arranged and signed between the UKAEA and the firms Rolls-Royce, Vickers-Armstrongs and Foster Wheeler during June and July 1956 detailing the services the UKAEA would provide to support their work. Anticipating Treasury approval, the Admiralty had written to Vickers-Armstrongs (Shipbuilders) to inform them of their intention to place an order with the company as main contractors and, where necessary, to use Rolls-Royce and Foster Wheeler as sub-contractors.[45] Treasury approval for HMS *Dreadnought* and the Dounreay Submarine Prototype was forthcoming in January with an agreed expenditure of £300,000 in 1956/1957. The Treasury advised that approval for the project was: '…without prejudice to any decision on an advance beyond the prototype of the nuclear power submarine'.[46] The Royal Navy's nuclear submarine project was once again getting the financial resources it required, the engineering problems were being overcome and Parliament was starting to take notice.

No debates over the Mark I enriched reactor were held in Parliament and only guarded references were made in Command Papers to investigating all possibilities for submarine propulsion.[47] From early 1956, however, the Government was becoming more open about the project. In his statement on the naval estimates, the First Lord, Lord Cilcennan, noted that scientists and naval officers had been working on the possibility of nuclear propulsion for some years: '…but it has only recently become possible to start practicable work of planning a marine power plant'.[48] This was the first explicit statement by a government Minister that the Admiralty was committed to developing a nuclear submarine and it was reported the following day in the major daily newspapers. *The Times* article noted that industry was also to work with the Admiralty on the project.[49] The article was tucked away on page eight suggesting that it was not a story of great importance; however, the convention of the day was for stories of international importance to be placed on the front page. As an example, HMS *Dreadnought*'s launch in 1960 was reported on page six, in contrast, the launch in 1967 of France's first nuclear-powered vessel, the ballistic missile

[45] TNA, AB 6/1051, Letter, CP.101/55 5, January 1956.

[46] TNA, T 225/1022, Letter, D.M.35/77/01, 6 January 1956.

[47] See, e.g., Admiralty statement of the first lord of the Admiralty. Explanatory of the navy estimates;1951–52 (Cmd.8476), p. 9.

[48] Parliamentary paper, Admiralty. Explanatory statement on the navy estimates 1956–57 by the first lord of the Admiralty; 1955–56 (Cmd.9697), p. 5.

[49] 'Atomic Ships planned', *The Times*, 24 February 1956, p. 8, Issue 53463.

submarine, *Le Redoutable*, made the front page of *The Times*.[50] In March, George Ward, Parliamentary Secretary to the Admiralty, advised the Commons that a subsidiary company was to be formed, Vickers Nuclear Engineering Limited (VNEL). Ward noted that he expected other companies to become associated with the project as it evolved.[51] VNEL was eventually formed in April 1956 and after protracted negotiations the contract was signed in June allowing work to start in earnest.[52]

Miracle Metals

In January 1956, the Admiralty formally wrote to the UKAEA requesting technical advice and assistance in pressurised water reactor technology. The UKAEA was advised that it was the Admiralty's intention to build a pressurised water reactor of approximately 80 MW output and commence trials on a land-based prototype by late 1959 and for trials in a submarine by 1962.[53] Professor Jack Edwards, the senior RNSS scientist in the Naval Section at Harwell, observed that the pressurised water reactor had been chosen as the most feasible reactor for installation in a submarine because of its compactness and good prospects of completing the project within a reasonable timescale.[54] Edwards also noted little was known in the UK of light water-moderated and water-cooled reactors; there were considerable gaps in the available knowledge and little information was available from the US. This was not unreasonable given the impact of the McMahon Act; however, British engineers adopted novel ways to collect information on nuclear energy, as noted in the previous chapter. Due to there being no official channel for the exchange of nuclear power generation information, as much knowledge as possible was gleaned from Congressional reports, US Atomic Energy Commission reports, press releases and other published documentation.[55]

[50] 'France Joins Nuclear Navies', *The Times*, 30 March 1967, p. 1, Issue 56902.

[51] Parliamentary debate, House of Commons sitting, 8 March 1956, 1955–56; Elizabeth II year 5; Columns 2279–2486, Fifth series Volume 549.

[52] DNP 2, NP184/2011, p. 17.

[53] TNA, AB 16/1764, Marine propulsion: naval nuclear submarine project; discussions with Admiralty, 1956, Letter, Lang to Peirson, 4 January 1956.

[54] Professor J. Edwards, 'Joint Panel on Nuclear Submarine Propulsion: Initial Problems of the Submarine Pressurized Water Reactor Design and the Related Experimental Programme', Lecture to the Institute of Marine Engineers, 23 January 1962.

[55] TNA, AB 6/1051, Paper, U.S.A. Nuclear propelled submarine, 19 May 1952.

Writing in 1956, however, Edwards observed that a great deal of experience had been obtained in the US on pressurised water reactors and was one: '…on which considerable data have been published.' It was also one of the systems that had been intensively investigated by the UKAEA and the Naval Section.[56] Rickover had committed to publishing a lot of information on pressurised water reactors that would be applicable to civil nuclear power stations; Cockcroft and Lord Plowden had received copies of *Corrosion and Wear Handbook for Water Cooled Reactors*, the data from which had been developed in the naval reactors programme and the Shippingport pressurised water reactor project.[57] The fuel cladding material, zircaloy, had initially been developed for the US submarine programme; however, it became evident there were also applications as a cladding material in civil pressurised water reactors, such as the reactor built at Shippingport, Pennsylvania: '…extensive data was declassified; for example a volume on the *Metallurgy of Zirconium* by Lustman and Kerne was published in 1955'.[58] Murray noted that it was implicit from the preface of Lustman and Kerne's book that Rickover had instigated its publication to encourage fundamental research on the metallurgy of zirconium and stimulate commercial development of zirconium. More pressurised water reactor nuclear information entered the public domain from the Shippingport reactor which impacted greatly on nuclear technology in general because Shippingport had no military applications: '…every aspect of its design and operation could be declassified'.[59] No doubt these and other publications would have been made available to engineers from the technical library at Harwell.

Robert Gordon, manager of Westinghouse Electric's Special Alloy Development Department, was chosen to: '…develop and manufacture the nuclear core of the Nautilus'.[60] Gordon noted that the design report

[56] J. Edwards, 'Nuclear Power Plants: An Outline of the Experimental Programme Necessary to Support a Nuclear Power Project', *Journal of the Royal Naval Scientific Service*, Vol. 11, (1956), 255–78 (p. 255).

[57] TNA, AB 6/1987, US/UK exchange of reports on nuclear marine propulsion, 1957-61, Letter, Rickover to Cockcroft, 30 July 1957.

[58] TNA, AB 6/2492, Admiralty Submarine Programme: organisation and review, 1960–65, Note, P. Murray, Naval Policy at Harwell, 15 May 1964.

[59] Hewlett and Duncan, *Nuclear Navy*, p. 254.

[60] Robert B. Gordon, 'Working for Admiral Rickover', (Naval Historical Foundation Memoir Program, Washington D.C., 2000), p.3. Gordon's memoirs were later published in 'Miracle metals serving in war and peace', *Power Engineering*, Vol. 104, (10), (Oct., 2000), 118–228.

specified using zirconium as the structural material to contain the fuel and using a chemically similar metal, hafnium, to control the power level. Gordon referred to these two materials as "miracle metals" due to their qualities and use in the reactor core (zirconium has excellent mechanical properties and a very low neutron capture, whereas hafnium is a very efficient neutron absorber). Both metals are found in the same ore and require chemical processes to separate the small percentage of hafnium impurity from the zirconium; with hafnium present zirconium is of no use as a reactor material. 'In the Fall of 1949, the world supply of these two (then rare) metals was measured in ounces and their purity level was very inadequate'.[61] In 1948, prices for zirconium varied between $135 and $235 per pound, by 1954 the purity and quality had risen, and the price had fallen to $13.10 per pound. By 1963, zirconium was considered a commercial product: '...falling in some cases below $4 per pound'.[62] In the UK a commercial plant was producing zirconium, the metal cost £10 per pound but contained between half and three percent hafnium. The ratio of zirconium to hafnium as an ore is roughly sixty to one and for zirconium to be employed in reactor construction, hafnium had to be reduced to less than 0.05 percent. To further reduce the hafnium content in the UK plant, it was estimated that the price of zirconium would increase to between £13 and £15 per pound.[63] The currency exchange rate in the 1950s was approximately $2.80 to £1.00, the cost to Britain buying US-produced zirconium would have been around £4.67 per pound, cheaper than the home-produced metal. The UKAEA had advised the Admiralty that they were confident of meeting any of their demands for zirconium from either British or foreign sources.[64] Economically it would make sense to purchase zirconium from the US, however, although it was recognised there would be a capital cost to increase domestic production of zirconium for a limited market, the submarine project, governments of the period were keen to avoid unnecessary dollar expenditure. The political implication of relying on foreign supplies of material for the nuclear submarine programme was also considered. President Dwight D. Eisenhower and some members of his administration were keen to help the UK's submarine programme, but their

[61] Gordon, 'Working for Admiral Rickover', p. 4.

[62] Hewlett and Duncan, *Nuclear Navy*, p. 289.

[63] TNA, AB 15/2186, Paper, Enriched Reactor Mk1 Feasibility Report – Section I, May 1952, Chapter 3, p. 3.

[64] TNA, AB 16/1379, Telegram, Hinton to Peirson, 22 July 1957.

political goodwill was countered by Congress. The Admiralty and the UKAEA needed reliability in the supply chain to meet their programme targets until such time when assistance from the US allowed the supply of such materials for use in reactor components.

A meeting was held to obtain an opinion on the best fuel and canning materials on which to concentrate the efforts of the Naval Section. Among the possibilities considered, the two favoured fuel elements were zirconium-canned uranium/zirconium fuels and mild-steel-canned uranium oxide fuels. Although zirconium and its alloys were considered a better proposition, production facilities were deemed inadequate. Opinion favoured a uranium oxide fuel element canned in mild or stainless steel, preferably mild steel.[65] The Admiralty was already conducting tests on different zirconium alloys, and the meeting concluded that nuclear calculations on uranium oxide/mild steel cermets and uranium oxide/zirconium cermets could start as soon as the current uranium/zirconium alloy work was completed. Work proceeded equally on mild steel and zirconium fuel elements until March 1957 when a decision was taken to proceed with uranium oxide/steel: '…at that time all work on uranium/zirconium ones was stopped in order to concentrate the effort'.[66] Some basic information on fuel and canning materials was exchanged with the Americans during a visit by Rickover and his party in August/September 1956; however, discussions were limited to the UKAEA's effort on gas-cooled reactors.[67] A report on the satisfactory reprocessing of a zirconium core, published by the USAEC and received by the Admiralty in October 1957, influenced the Admiralty to stop production of the uranium oxide/steel fuel elements and to change to uranium/zirconium fuel elements, even though this meant a further delay in the project of six months.[68] This change was to be an important decision as we shall see.

The availability of hafnium was a problem for most of the nuclear submarine programme. Rear Admiral Nuclear Propulsion, G. A. M. Wilson, wrote to Strath noting some short-comings with the Exchange Agreement. The restriction on information concerning absorber materials was of

[65] TNA, AB 6/1051, Note, J. Edwards, 25 November 1955.

[66] TNA, AB 16/1380, Marine Propulsion: naval reactor for Admiralty submarine, 1957–62, Telegram, Stewart to Strath, 21 October 1957.

[67] TNA, AB 15/5939, Report on the visit of Admiral Rickover and party to the Metallurgy Division AERE, 1956.

[68] TNA, ADM 1/26740, Fuel elements for nuclear submarine: Admiralty Board decision, 1957, Letter, Lang to Strath, 24 October 1957.

concern as the project was practically forced to use an alternative to hafnium due to no information on alternatives being available from Westinghouse. However, in Wilson's covering letter he describes how two Rolls-Royce personnel had had a conversation with staff from Westinghouse about the lack of information concerning absorber material; their reply was to advise Rolls-Royce to write to them: '...because all this information has been de-classified anyway'.[69] Hafnium is about thirty percent more effective than other neutron absorbers, such as cadmium, which is why it was the preferred material for fabricating the control rods.[70] Recollecting events some forty years later, Edwards described how the project had preferred hafnium because of its advantageous nuclear properties but there was none in the UK available for testing. 'Somehow or another when this became known, a small piece of reactor grade hafnium came into our hands during the American trip...'[71] Edwards explains how this small sample was invaluable, being put through various tests in the UK, allowing the project to formulate their future design plans. Nonetheless, concerns were still being expressed during 1958. Edwards wrote that an order had been placed for twelve tons of hafnium-free zirconium for the year ending March 1958 which would have yielded some 800 pounds of hafnium oxide. Edwards calculated another 120 pounds would have accumulated from other sources which would not be enough to satisfy the core requirements.[72]

At a meeting to discuss progress with the nuclear submarine Wilson pointed out the difficulty in obtaining hafnium, Sir William Cook, head of the RNSS, suggested getting an undertaking from the Americans to supply. Harrison-Smith advised that the Americans: '...had barely enough for their own requirements'.[73] However, Macmillan's objective of achieving a comprehensive nuclear exchange treaty, interdependence with the Americans, was slowly coming to fruition. Macmillan's main foreign policy was to pursue the London-Washington special relationship as a priority over Europe and with it to prevent war by developing an independent

[69] TNA, AB 16/1380, Letter, Wilson to Strath, 13 November 1957.

[70] A. L. Kendrick, 'The R.N.S.S. and Neptune Part 1', *Journal of the Royal Navy Scientific Service*, Vol. 14, (1959), 226–36 (p. 235).

[71] Private correspondence, Jack Edwards to Vice Admiral Sir Robert Hill, 6 May 1998.

[72] Edwards, Joint Panel on Nuclear Marine Propulsion, p. 19.

[73] TNA, DEFE 69/749, Dreadnought: negotiation of purchase of US propulsion plant, 1957–58, Minutes, 27 March 1958, item V.

nuclear deterrent.[74] Achieving nuclear interdependence with the US would be a means of sharing rather than duplicating nuclear knowledge, following a meeting of atomic experts Macmillan noted that in some respects the UK was as far ahead and even further advanced in the art than the Americans who thought the interchange of information would be all give.[75] Indeed, the US encouraged the UK to complete its "Grapple" tests which were of great interest to them.[76] Interdependence would be a chance for the Admiralty to alleviate many of the problems in their nuclear submarine programme.

In the spring of 1958, negotiations were proceeding with an amendment to the 1956 Civil Exchange Agreement which would culminate in the 1958 Mutual Defence Agreement. The Admiralty was keen to ensure that the future exchange agreement not only covered uranium requirements but also hafnium. The Foreign Office contacted the Embassy in Washington to ensure that the amended bilateral agreement was worded sufficiently to cover not only enriched uranium but also other special nuclear materials such as hafnium.[77] The problems concerning availability of hafnium were finally resolved by the Mutual Defence Agreement which enabled the UK to buy a complete submarine reactor plant from the US. The Admiralty also placed a contract, through Rolls-Royce, with ICI to procure enough hafnium for the Dounreay Submarine Prototype, although discussions concerning the supply of hafnium continued for a couple more years.[78]

The Fuel Element Decision

Following the Technical Mission visit to the US, the Admiralty wanted to review the decision to use steel as the material for fabricating fuel elements based on American experience, indicating that zirconium may be a more suitable material for fabricating fuel elements.[79] At a meeting in July 1957,

[74] Alistair Horne, *Macmillan 1957–1986: Volume II of the Official Biography* (London, Macmillan, 1989), p. 22.

[75] Harold Macmillan, *Riding the Storm 1956–1959* (London, Macmillan, 1971), p. 565.

[76] Note: Operation Grapple was the UK Hydrogen Bomb tests conducted in 1957–1958.

[77] TNA, PREM 11/2554, UK-US co-operation in the field of atomic energy, 1957–58, Telegram, No. 3226, 30 May 1958.

[78] TNA, T 225/1193, Development of a nuclear submarine, including purchase of machinery from USA [...], 1959, Letter, Mawer to Hall, 12 June 1959.

[79] TNA, AB 16/1379, Letter, GFI/MAT.25383/57, 15 July, 1957.

Rear Admiral Wilson considered the choice of materials for the fuel elements. Wilson noted it had been decided previously to proceed with uranium oxide/steel elements, partly on the grounds of doubts on the supply of British-refined zirconium which made the expense of full-scale development of a material for a limited application difficult to justify. Partly, also, it was believed that there was a better long-term development potential for uranium oxide/steel elements. The UKAEA had advised that the use of zirconium added serious complexity to the proposed chemical separation plant for treating irradiated fuel elements, which would add £400,000 to the capital cost and a further £250,000 to increase British production of zirconium to meet Admiralty requirements.[80] Despite the UKAEA's reservations, following the recent Technical Mission visit to the US there was a revived interest in the zirconium core, the main reasons for this was a more favourable supply of zirconium that could be purchased from the US or Japan and the fact that: 'A zirconium core uses only a little over half the amount of fissile fuel required by the steel core'.[81] A smaller reactor core design could be achieved, and the price of zirconium was likely to give an overall saving. It was estimated that the total fuel costs for the submarine programme up to 1970 would be roughly £23M for steel cores and £16M for zirconium cores. It was noted that economy of operation of either core depended on the value of fissile material ultimately recovered and returned to stock. 'If reprocessing of Zirconium fuel elements is practicable the overall economy will favour this type of core'.[82] Wilson argued that given the technical, economic and political considerations, there was little to choose between the fuel elements and the submarine's completion date was unlikely to be affected by the choice. The recommendation was to proceed with a reactor designed with uranium oxide/steel elements. However, it was thought necessary to mention the reduction in fissile fuel required for a zirconium core if satisfactory reprocessing could be developed. The recommendation was accepted by Admiralty's finance section in late August and the Admiralty notified the UKAEA that the reactor design should proceed on the basis of uranium oxide/steel fuel elements.[83]

However, a fortnight later Wilson noted that the recent reservation expressed by the UKAEA concerning the reprocessing of zirconium cores

[80] TNA, ADM 1/26740, Paper, RANP/23/8, 30 July 1957, p. 4.
[81] TNA, ADM 1/26740, Paper, RANP/23/8, 30 July 1957, p. 1.
[82] TNA, ADM 1/26740, Paper, RANP/23/8, 30 July 1957, p. 5.
[83] TNA, AB 16/1379, Letter, G.F.513/57, 19 September 1957.

was unfounded as the US had recently reprocessed their first zirconium core.[84] Wilson advised the meeting that they should now proceed with the zirconium core and accept a delay of some six months to HMS *Dreadnought*'s completion date. A decision on Wilson's view does not appear in the records; however, Harwell's Head of Metallurgy, H. M. Finniston, was certainly aware of the Admiralty's intention to change to zirconium elements by mid-October and advised the UKAEA Secretary, D. E. H. Peirson, that he envisaged a delay of six to twelve months.[85] Sir John Lang, Secretary of the Navy, wrote to Strath to inform him that, knowing zirconium could be reprocessed safely and that the methods of doing so are within the scope of the exchange agreement: '…we assume that the A.E.A will feel assured of their ability to reprocess zirconium elements for us'.[86] The basis for the Admiralty's decision to proceed with zirconium-based fuel elements for HMS *Dreadnought* is open to conjecture. The meeting at which Wilson advised changing to zirconium also discussed the threat to the Dreadnought project posed by possible political support for a nuclear-powered tanker. Wilson had written an earlier paper stating the case for a Fleet Replenishment Tanker for the Galbraith Committee.[87] In the paper Wilson argued that machinery developed for the Fleet Tanker would be applicable to a merchant ship.[88] Mountbatten told the meeting he was certain Wilson supported HMS *Dreadnought* but: '…the figures that the latter had produced in his paper had made it very easy to argue that it would be better from every point of view to build the tanker rather than the submarine'.[89]

Although the tanker was a paper study and some three years behind the Dreadnought project, Mountbatten was concerned that once people became aware of it there would be support for the tanker especially at the Air Ministry. Considering the first Polaris submarine, the USS *George Washington*, was laid down in 1958, the fact that the UK was still pursuing a policy of what Macmillan called the "Great Prize" of nuclear

[84] Broadlands Archive, MB1/I397, Sea Lords' Meeting 1957–1959, 8 October 1957, Item 1.

[85] TNA, AB 16/1380, Letter, Finniston to Peirson, 18 October 1957.

[86] TNA, ADM 1/26740, Letter, Lang to Strath, 24 October 1957.

[87] Full title: First Lord's Committee on the application of nuclear power to marine purposes.

[88] DNP 1, ALO 02617/97 – 000378, Paper, MAT.25357/57, 27 June 1957.

[89] Broadlands Archive, MB1/I397, Sea Lords' Meeting 1957–1959, 8 October 1957, Item 1.

interdependence with the US, and that the Polaris Sales Agreement lay six years in the future, it was quite prescient of Mountbatten to tell the meeting that to adopt the tanker in place of HMS *Dreadnought*: '…would suit the Air Ministry well, as not only would it remove a rival to the deterrent, but the tanker would require less fissile material than the *Dreadnought*'.[90] Given these facts it seems entirely probable that the Admiralty adopted the uranium/zirconium element on the grounds of reducing by nearly half the fissile material required for HMS *Dreadnought* and, thus, countering any Air Ministry argument that may have been used to make a case for cancelling the nuclear submarine programme in favour of the tanker. During his brief tenure as Minister of Defence, Macmillan had acknowledged that the Armed Forces were still bloated from World War II and had not adjusted fast enough. Shortly after the Suez crisis, whilst still Chancellor, Macmillan wrote a memorandum to Anthony Head, Minister of Defence, arguing the case for radical cuts in defence spending.[91] In the memorandum, Macmillan prioritised the prevention of global war; this would be achieved by nuclear deterrence, atomic weapons carried by the Royal Air Force and a ballistic missile, to be developed.

On becoming Prime Minister in January 1957, Macmillan appointed Duncan Sandys as the Minister of Defence with a brief to cut defence spending. With the Royal Air Force having the major role, perhaps it was natural that the nuclear submarine programme would come under scrutiny. Although the axe was to fall on the Army and the Royal Navy it is reported that Sandys developed a good relationship with Mountbatten.[92] This is no surprise as Mountbatten immediately set out on a charm offensive as only he could; in March, Sandys accepted an invitation to dine at Mountbatten's London residence where the Queen and Prince Philip would be guests.[93] In October, Sandys, with other senior Admiralty figures, would spend a day at sea in the USS *Nautilus* organised by Mountbatten, and in November Sandys would enjoy a Trafalgar Night

[90] Broadlands Archive, MB1/I397, Sea Lords' Meeting 1957–1959, 8 October 1957, Item 1.

[91] Horne, *Macmillan 1957–1986: Volume II*, p. 45.

[92] ODNB, Piers Ludlow, 'Sandys, (Edwin) Duncan, Baron Duncan-Sandys (1908–1987)', *Oxford Dictionary of National Biography*, Oxford University Press January 03, 2008. http://oxforddnb.com/view/10.1093/ref:odnb/9780198614128.001.0001/odnb-9780198614128-e-39858 [Accessed 17 August 2018].

[93] Broadlands Archive, MB1/I391, Duncan Sandys, PC, MP, Minister of Defence 1957–59, Letter, Sandys to Mountbatten, 11 March 1957.

dinner onboard HMS *Victory*.[94] Sir Rhoderick McGrigor, Mountbatten's predecessor as First Sea Lord, had written to Mountbatten stating that he was: '...somewhat horrified at the news of Duncan Sandys becoming Defence Minister', ending the letter: 'All my best wishes to you in the struggle ahead'.[95] Sandys was a great advocate for basing the UK's defence on the policy of nuclear deterrence and saw the RAF as the prime carrier of the deterrent in his Defence White Paper, *Outline of Future Policy*, issued in April 1957. The Paper envisaged a huge reduction in the Army with the ending of conscription; the Royal Navy's role was defined merely as: 'Peace time emergencies or limited hostilities'.[96] Indeed, there was no perceived role for the Royal Navy in total war as it was assumed at that time that any conflict between the NATO member countries and the Warsaw Pact would immediately go nuclear. This was a time before the proliferation of hydrogen weapons and the realisation that the use of a tactical nuclear weapon would lead to greater use of nuclear weapons as a defence. This scenario eventually led to the adoption of a strategy known as Mutually Assured Destruction (MAD) which evolved in the early 1960s with a second strike capability guaranteed by the opposing belligerents' SSBNs. Considering at that time many military commanders and politicians expected a future war with the Soviet Union would be short and intense with air power playing the major role, Eric Grove raised the question which was possibly at the back of Sandys' mind: 'Was it really prudent to re-arm for a long-drawn-out World War II type war under the new conditions?'[97]

After Sandys' White Paper, each of the services gave a series of briefings to back benchers and the press to either promote one area of interest or prevent Treasury cuts in another area. The RAF went about their briefings with alacrity and mounted a three-day conference for selected RAF officers and Air Ministry officials, a further one-day conference was held 6 May 1958 for selected defence correspondents.[98] Laurence Martin argued that at these briefings, the deterrence operated by the RAF was held as the only realistic policy for the UK. During the mid-1950s, the "V" bomber force was being built to deliver the deterrent which was initially based on

[94] Broadlands Archive, MB1/I391, Letter, Sandys to Mountbatten, 25 November 1957.
[95] Broadlands Archive, MB1/I391, Letter, McGrigor to Mountbatten, 14 January 1957.
[96] Parliamentary paper, Defence Outline of Future Policy; 1956–57 (Cmnd.124), p. 6.
[97] Grove, *Vanguard to Trident*, p. 81.
[98] Martin, 'The Market for Strategic Ideas in Britain: The "Sandys Era"', p. 32.

Blue Danube, a gravity bomb. An improved bomb codenamed *Red Beard* was also a gravity bomb and this left the RAF bombers exposed to attack from improved Soviet air defences. The UK had developed its own air-launched stand-off missile, *Blue Steel*, to maintain the "V" bomber force as a credible deterrent but was already outdated when it entered service in 1963, a Mark II version was cancelled in 1960 due to development difficulties. From 1960 the UK sought to purchase the US stand-off missile, *Skybolt*, which was under development to improve the American deterrent. However, the US Government cancelled the *Skybolt* missile system in 1962, also due to development difficulties, much to the dismay of the British Government.

Development of the Polaris system began in 1956 and was not taken seriously by the British Government; Martin noted the influence of the USAF in providing the RAF and Ministry of Defence with sceptical assessments of its performance.[99] There is no doubt that the Air Ministry would use these briefings to counter any future Admiralty talks on adopting the Polaris system. In contrast to the RAF, Martin argued that the Royal Navy was more restrained and that it saw its role in deterrence in terms of the aircraft carrier; however, the aircraft carrier was also criticised for being vulnerable to Soviet anti-ship missiles. Martin also noted that: '...the Navy was slow to develop an alternative interest in the Polaris'.[100] However, it can be argued that through being discreet in their briefings, the Navy would make a shrewd case for a sea-based deterrence founded on technical and engineering fact once the US Navy had deployed their own Polaris submarines. Mountbatten had discussed Polaris in 1957 with his counterpart, Admiral Arleigh Burke USN, when Burke ordered the programme to proceed.[101] Indeed, confiding to a USN Admiral some years later Mountbatten confirmed that a Royal Naval officer missile expert was seconded to the US Polaris programme from the outset.[102] It would appear that Mountbatten wanted to make sure that when the Royal Navy had a proposal for a sea-based deterrent they could deliver it to Cabinet Ministers confident in the knowledge that both Polaris and nuclear propulsion had already been proven as dependable technologies and that moving to a

[99] Martin, 'The Market for Strategic Ideas in Britain: The "Sandys Era"', p. 28.
[100] Martin, 'The Market for Strategic Ideas in Britain: The "Sandys Era"', p. 32.
[101] Broadlands Archive, MB1/I300, Quarterly Newsletter, 1 August 1957, p. 11.
[102] Broadlands Archive, MB1/J40, Letter, Mountbatten to Rear Admiral I. J. Galantin, 21 January 1965.

sea-based deterrence was the best means of defence as it created uncertainty for the Soviets from where a retaliatory attack could be launched. Mountbatten certainly had an eye on a future role for the Royal Navy not envisaged in 1957 by Sandys nor the Ministry of Defence, that of carrying a sea-borne deterrence. Mountbatten wrote: '…I hope we shall now have his [Sandys] wholehearted support for the *Dreadnought* and eventually the *Polaris*-type nuclear submarine'.[103] However, to carry a submarine-based nuclear deterrent, it was imperative that nuclear propulsion be mastered in a timely manner.

Neptune: The Zero Energy Experimental Reactor

For nuclear propulsion to be mastered, basic reactor physics information relating to light water-cooled/moderated, highly enriched uranium systems were required to guide the design for the first submarine reactor. The Admiralty's zero energy experimental reactor, Neptune, was conceived in early 1956 to assist: '…in resolving major uncertainties in many factors influencing the calculation of criticality, control, temperature effects on reactivity, and endurance for the power reactor'.[104] Rolls-Royce was contracted to design and build the reactor at Harwell, the AERE would be responsible for the physics programme and the buildings. The basic design specifications were agreed in July 1956; details of the numbers and grades of additional staff were forwarded to the AERE and fissile material was allocated by the Chiefs of Staff for use during 1956/1957 to support Neptune.[105] It has been noted earlier that the UKAEA was eager to involve private industry in developing nuclear power; to support Neptune the Authority entered into enabling contracts with Vickers-Armstrongs, Rolls-Royce and Foster Wheeler during the summer of 1956. Committees consisting of members of the relevant interested parties were formed to coordinate the experimental work and ensure there was no duplication of effort (Fig. 4.1).[106]

Although fissile material had been allocated for 1956/1957, the Admiralty had not apportioned money to purchase the fissile material. A

[103] Broadlands Archive, MB1/I300, Quarterly Newsletter, 1 November 1957, p. 9.

[104] J. Edwards and J. E. R. Holmes, 'Neptune: The Admiralty Zero Energy reactor', *Journal of the Royal Navy Scientific Service*, Vol. 13, (1958), 223–32 (p. 223).

[105] TNA, AB 6/1051, Letter, Hudspith to Synott, 2 May 1956.

[106] TNA, AB 6/1051, Note, Edwards, 6 June 1956.

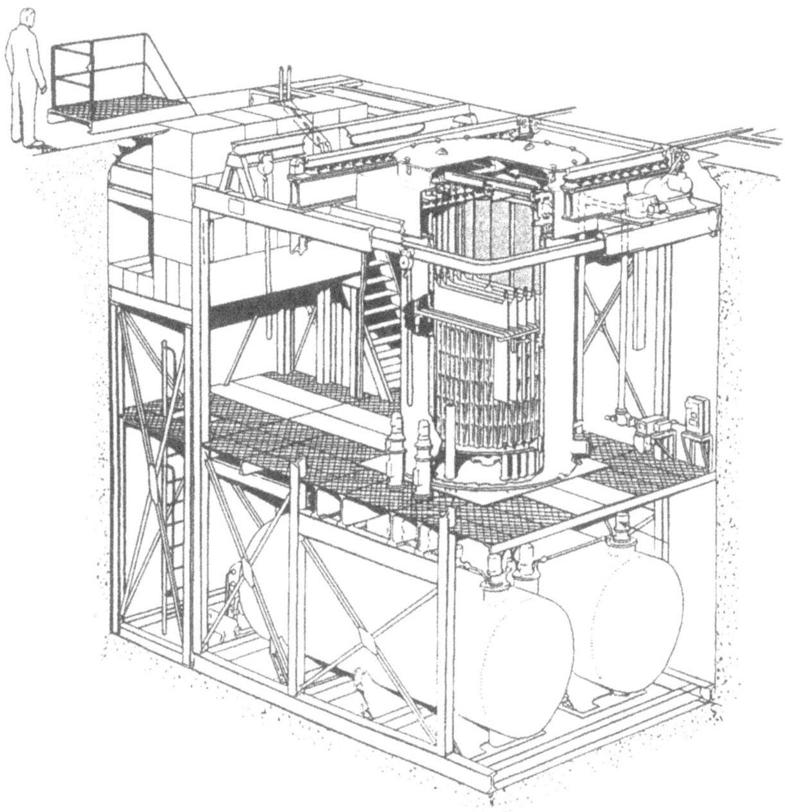

Fig. 4.1 Neptune reactor diagram

provision of £298,000 in the 1956/1957 Estimates had been made for all nuclear propulsion services. Under a formula sanctioned by the Treasury, the UKAEA estimated the cost to the Admiralty of the fissile material at roughly £1,000,000 with a further £50,000 fabrication costs. The Admiralty sought to purchase the material outright during the current financial year as a cheaper option to hiring as suggested by the UKAEA.[107] The Treasury replied agreeing to the Admiralty purchasing the fissile material during the current year but would not commit the Treasury in

[107] TNA, T 225/1022, Letter, G.F.5016/56, 13 June 1956.

agreeing to a Supplementary Estimate being submitted. The Treasury also advised that the nuclear submarine project raised wider issues and posed a number of questions that needed to be discussed, such as future requirements for fissile material. The Treasury also questioned the decision to concentrate effort on a nuclear submarine asking whether it was likely to unearth: '...special problems which would not arise for surface vessels and so slow down the rate of research and development in this field?'.[108] As previously noted, the Admiralty had requested the Treasury to sanction spending during 1956/1957 to develop a prototype set of machinery for a nuclear propelled submarine by entering into a contract with Vickers-Armstrongs.[109] Although the Treasury gave approval to proceed with the prototype in January 1956, it appears that they were now requesting the Admiralty to justify the development of the submarine and is perhaps political evidence of attempts to move the development project to the nuclear tanker proposition.

Estimates of the cost of fissile fuel were overoptimistic during the 1950s, possibly as an attempt to secure additional government investment into nuclear propulsion for the merchant marine. In a memorandum on the Economics and Logistics of Nuclear Propulsion for the Galbraith Committee, the cost of fuel oil to the navy in 1956 was quoted as 0.53 pence per shaft horse power hour (SHP/Hour). The high cost of fissile material and the elementary reactor under development would achieve a cost of 4d per SHP/Hour. Captain Sarell argued that by 1965 the lower price of fissile material combined with more advanced practice should produce a cost of 1.5 pence per SHP/Hour. This cost assumed enriched uranium priced at £12,000 per kilo; however, if the price fell to £9000 per kilo as quoted by the UKAEA, the cost would fall to 1.2 pence per SHP/Hour.[110] It is little wonder with figures illustrating increasing fuel oil costs in comparison with declining nuclear fuel costs that the Treasury was looking to justify spending the research and development costs allocated to the submarine project on the merchant marine which would be of more benefit to the country.

The Admiralty replied to the Treasury's request for more information by giving a broad outline of the project from designing the test site and facilities in January 1957 through to commissioning trials of the

[108] TNA, T 225/1022, Letter, Serpell to Samuel, 7 August 1956.
[109] TNA, T 225/1022, Letter, McKinnell to Radice, 8 December 1955.
[110] DNP 1, ALO 02617/97 – 000378, Memorandum, 035, 14 November 1956.

submarine in December 1961. There was a prospect of the US releasing information on propulsion reactors under the 1956 Civil Exchange Agreement; however, the Admiralty advised that even if information was received from the US they would still have to conduct zero energy experiments for safety reasons. The Admiralty argued that it was impossible to reproduce exactly the materials used in the US and it was therefore important to prove their own materials, adding that there was no alternative but to continue with the programme: '…any U.S. information which we obtain may well result in a slight reduction of our own work, it cannot replace it'.[111] Construction progressed on Neptune, which was planned for completion in July 1957; however, Rolls-Royce had sub-contracted Neptune's erection to another firm whom the AERE had little confidence in keeping to the programme.[112] Although Neptune was viewed at the time as a bottleneck in the programme, the timescale from agreement of the design to the reactor going critical was a little over sixteen months. Costing £233,000, Neptune, the Royal Navy's first nuclear reactor, went critical for the first time at 8:30 PM 7 November 1957.[113]

Neptune was devised to operate at one or two watts of power and rarely exceeded ten watts, hence the handling difficulties encountered with "hot" fuel elements was not a problem. The design of Neptune allowed for fuel elements and control rods to be quickly assembled in a variety of simple configurations to enable calculations to be made and compared with other core layouts.[114] Evidence from these calculations would determine the critical size of the reactor and allow for the most favourable arrangement of the fuel elements and control rods. Neptune was one of a variety of reactor physics experiments in progress and it was envisaged to take six months of experiments with Neptune before all reactor physics information would be available, which were required to finalise the core and fuel parameters. Edwards noted that all the early work in Neptune was conducted on fuel arrangements corresponding to highly enriched uranium oxide/steel elements even though the decision to change to uranium/zirconium had been taken in October 1957. Edwards argued that the zirconium plates for Neptune that had been ordered as an insurance policy in early 1957 had yet to be fabricated and whilst waiting

[111] TNA, T 225/1022, Letter, G.F.I/G.F.499/56, 18 September 1956.
[112] TNA, AB 16/1379, Note, A.E.R.E. D/M310, 6 February 1957.
[113] DNP 2, NP184/2011, p. 20.
[114] Edwards and Holmes, 'Neptune:', p. 224.

for their delivery it seemed wrong to leave such work unfinished after so much had been achieved.[115] Neptune operated throughout 1958 and was due to cease operation in June 1959. After it was shut down, Neptune was dismantled and kept in storage until a decision would be made on its future.[116] Neptune would eventually be moved to Rolls-Royce's premises at Raynesway, Derby, to continue nuclear physics experiments on future advanced cores for the Royal Navy.

By the end of 1957, the Admiralty's nuclear submarine programme had reached a critical stage. Rear Admiral G. A. M. Wilson had been appointed Rear Admiral Nuclear Propulsion to co-ordinate the nuclear engineering side of E-in-C's office and, to give a stronger management structure, the Dreadnought Project Team was formed under the leadership of Rowland Baker RCNC at Bath. Contracts had been placed for the hull and machinery with Vickers-Armstrongs and the UK's reactor design and fuel elements had been agreed upon with enabling contracts established between the main sub-contractors and the AERE. The programme was proceeding as expected and the Queen had approved the name *Dreadnought* for the first nuclear-powered submarine. However, two unrelated events in 1957, months apart, would combine to change the course of the programme, and of the Royal Navy's own nuclear reactor design.

With Eisenhower's approval, American officials were already secretly dealing with members of Eden's Cabinet, including Macmillan, when, as a result of the Suez Crisis, Macmillan succeeded Eden as prime minister in January 1957.[117] Due to their personal wartime relationship, within days the Americans proposed a meeting in Bermuda between Eisenhower and Macmillan which would attempt to put the strain between the two countries over Suez behind them. Macmillan's main foreign policy objective of securing nuclear interdependence with the Americans was something that Eisenhower agreed with, but had his own political difficulties with Congress. However, Macmillan's chance to change US political opinion came in October when the Soviets launched the first successful satellite, *Sputnik*. The launch of *Sputnik* on 4 October 1957 had caught the Americans unawares and caused panic in some quarters as it appeared that the Soviet Union not only had atomic and hydrogen bombs, but had

[115] Edwards, Joint Panel on Nuclear Marine Propulsion, p. 8.

[116] TNA, AB 6/1818, Liaison with the Admiralty, 1956–61, Note, J. Smith, 10 June 1959.

[117] Chester J. Pach & Elmo Richardson, *The Presidency of Dwight D. Eisenhower, Revised Edition* (Lawrence, University of Kansas Press, 1991), p. 135.

developed a rocket with the capability of delivering them to the American continent. Macmillan wrote to the Secretary of State, John Dulles, suggesting the time was right for pooling defence resources but refrained from mentioning the McMahon Act at that time.[118] Three weeks later, at a meeting in the White House, Eisenhower surprised the British by producing a directive dealing with nuclear collaboration between the two countries. In effect, it was the end of the McMahon Act which Eisenhower described to Macmillan as: '...one of the most deplorable incidents in American history, of which he personally felt ashamed'.[119] Macmillan had secured the key to his major international policy objective of nuclear interdependence with the US; the "special relationship", as it is known, was about to begin.

By 1957, the Royal Navy was forging ahead with their nuclear submarine project, although there were concerns over hafnium supplies, most of the engineering problems had been solved and the programme seemed to be on schedule for a submarine to be at sea by late 1961. The US Navy was slowly building up their fleet, with three of the *Skate* class being launched in 1957; the Soviets too, would launch their first nuclear-powered submarine that year. Although there were still no political debates over the UK's nuclear submarine project, the coming of the "special relationship" was to have consequences for the Admiralty nuclear propulsion programme, and HMS *Dreadnought* in particular.

[118] Macmillan, *Riding the Storm*, p. 314.
[119] Horne, *Macmillan 1957–1986: Volume II*, p. 57.

CHAPTER 5

HMS/m *Dreadnought*

The importance of Admiral Hyman G. Rickover USN and his association with the Royal Navy's nuclear propulsion programme is not that well known; Rickover is universally acknowledged as the "Father of the Nuclear Navy" and was renowned for his brusque manner and uncompromising attitude towards the highest quality standards. Rickover's management skill in overseeing the design, manufacture and operation of the world's first power reactor is well recognised throughout the nuclear industry. It was these attributes, and his personal rapport with Mountbatten, that, as we shall see, contributed to the Royal Navy's nuclear propulsion development programme by imposing the same exacting standards to UK industry as was applied to US industry.

This chapter will also investigate the possible savings to independent research in the UK by receiving US nuclear technology information, and Rickover's offer to the UK to purchase a US propulsion reactor as a means to expedite the Admiralty's programme. Anomalies have been uncovered in Philip Ziegler's official biography of Mountbatten and in the Broadlands Archive appertaining to reference MB1/K208A. This was a draft chapter of Mountbatten's involvement in the nuclear propulsion programme which was not scrutinised after his untimely death. These anomalies are addressed in this chapter merely for future reference. Finally, the chapter will explore the apparent mismanagement of the Dreadnought project and

G. M. Jones, *The Development of Nuclear Propulsion in the Royal Navy, 1946–1975*, Security, Conflict and Cooperation in the Contemporary World,
https://doi.org/10.1007/978-3-031-05129-6_5

the adoption of a unified project team under the management of one leader responsible for the whole programme without recourse to the usual Admiralty departmental hierarchy. Overall, the chapter will look at the personal relationships that supported Macmillan's foreign policy objective of interdependence with the US and helped to cement what has become known as "The Special Relationship".

From the mid-1950s, Rickover had been in contact with representatives of the UK nuclear industry. Initially, his contact was with the AERE in relation to the civil nuclear programme. Indeed, it is worth noting that in April 1955, during construction of the Shippingport reactor, Rickover made a trans-Atlantic call to a British metallurgist to clarify a technical point.[1] Although the metallurgist is not mentioned by name, Sir Harold (Monty) Finniston was the chief metallurgist at Harwell, and he and his wife became good friends with Rickover and his wife. Finniston also hosted Rickover on one of his visits to the UK; this close relationship, which is unusual given Rickover's temperament and his treatment of other UK nationals, would suggest that Finniston was the metallurgist referred to by Hewlett and Duncan.[2] Rickover's later contacts were directed more towards the Admiralty and those involved with the Admiralty's nuclear submarine programme. Rickover began exploring the state of research and development in the UK, looking at the management of the programme, the budget allocated by the Government and the condition of industry and their ability to get involved with the demands and standards imposed by nuclear engineering. Rickover's views were paramount to the US Government on whether the UK had the technical competency and the capacity to benefit from their offer to purchase outright a naval propulsion reactor. It was Rickover's authority and advice that would be the deciding factor on the future of the Admiralty's reactor design and manufacture, and it would fall to Admiral of the Fleet, Earl Mountbatten of Burma, as First Sea Lord to sway Rickover's influence in the Royal Navy's favour.

With the signing of the Mutual Defence Agreement on 3 July 1958, the Admiralty's Naval Nuclear Propulsion Programme (NNPP) changed dramatically. Article III of the Agreement authorised the purchase of a S5W submarine nuclear propulsion plant, the US Navy's latest and most powerful reactor which was being fitted into the *Skipjack* class SSNs and the *George Washington* class SSBN submarines of their nascent Polaris programme. The S5W purchase also included the complete propulsion

[1] Hewlett and Duncan, *Nuclear Navy*, p. 246.
[2] Private correspondence, Michael Finniston, 12 January 2017.

machinery set and auxiliary equipment, the Agreement likewise allowed for the transfer of enriched uranium to fuel the plant and the offer to reprocess spent fuel in the US. There was, however, a price, not only monetary but knowledge-wise. Article III of the Agreement was limited to ten years, after which there would be no more transfer of nuclear propulsion technology; the shutters would come down isolating the Admiralty from any further reactor technology developed, or experience gained in the US. Indeed, the transfer of nuclear information was further limited by the Westinghouse/Rolls-Royce contract. In effect, the purchase of the S5W would give the Admiralty a datum from which to develop its own design and advance its own reactor technology independently of the US. The Admiralty would quickly have to learn how to stand on its own two feet.

Not everyone in the UK propulsion programme was happy with the 1958 Agreement and questions were asked concerning how much information should be requested and what kind of technology should be transferred. It was even questioned whether any of the technology and information was required given that the Admiralty's own nuclear programme was, by mid-1958, so far advanced. It was certainly recognised that accepting US assistance would come with restrictions and would limit the Admiralty's own programme. However, the Royal Navy's priority was to have a nuclear-powered submarine at sea, and as soon as possible. But for that to happen, measures would have to be taken to change what can best be described as a haphazard management system into a singular project team headed by one man with overall responsibility.

Admiral Hyman G. Rickover USN

There is a consensus amongst historians on what has been written about the UK's nuclear propulsion programme that during the 1950s Rickover made three "official" visits to the UK. Rickover's first visit is recorded to have taken place 20 August to 4 September 1956.[3] Rickover's second visit took place 20–22 May 1957, and his third visit 24 January 1958. Rickover's first visit followed closely after the 1956 amendment to the 1955 Act allowing for cooperation on the civil uses of atomic energy, the Act was also intended to allow the transfer of nuclear propulsion information. Most of Rickover's visit was confined to the AERE, Harwell, where discussions centred on chemistry and metallurgy, although he also visited Calder Hall and

[3] TNA, AB 15/5939, Report on the visit of Admiral Rickover, 1956.

the UKAEA Industrial Group based at Risley.[4] Rickover's first day in the UK was spent at *Broadlands*, Mountbatten's country residence in Hampshire. Rickover talked openly and when Mountbatten asked his advice on what the Admiralty's next course of action should be Rickover replied that the Admiralty should buy a reactor from Westinghouse.[5] This was not the first time Rickover had proposed this line of action, however, he was speaking "off record" and nothing developed from the talks.

The 1956 amendment came into force 15 June and allowed for the exchange of restricted nuclear information applicable to civil uses; however, the Act also allowed for information relating to the propulsion of naval vessels to be exchanged: '…to the extent and by such means agreed'.[6] It was foreseen that information on military reactors would be exchanged through the UKAEA representative in Washington who would then disseminate the information back to the relevant departments in the UK.[7] However, the UK Government, and the Admiralty in particular, were disappointed with this amendment due to US political infighting. President Eisenhower had been championing closer nuclear ties with the UK since the start of his Presidency but was thwarted by Congress against his minority administration. The Pentagon had sent a draft of the 1956 amendment to the USAEC, an agency of the US Government established by the McMahon Act of 1946, which transferred control of atomic energy from military to civil supervision. However, the USAEC was overseen by Congress through the Joint Committee on Atomic Energy (JCAE), and the Pentagon had omitted to inform the JCAE. Upon discovering that they had not been consulted, the JCAE insisted that the Eisenhower administration suspend the new Agreement. 'As a minority President, and faced with an imminent election, Eisenhower had little choice but to agree'.[8] However, the previous February the Defense Department had asked the Attorney General, Herbert Brownell, for his confidential

[4] TNA, DEFE 7/2055, Telegram, PEW 1953, 28 February 1957.

[5] TNA, AB 6/1051, Letter, Edwards to Cockcroft, 5 September 1956, (enclosed Minute, 20 August 1956).

[6] Parliamentary paper, United States No. 2 (1956) Amendment to agreement between the government of the United Kingdom of Great Britain and Northern Ireland and the government of the United States of America for co-operation on the civil uses of atomic energy of June 15, 1955; 1955–56 (Cmd. 9789), Article I *bis*.

[7] TNA, AB 16/3162, Nuclear Powered Submarines; Exchange of Information with the United States of America, 1956–57, Letter, FB/270/56, 30 July 1956.

[8] Baylis, 'The 1958 Anglo-American Mutual Defence Agreement:', pp. 431–432.

opinion on whether it was legal, under the Atomic Energy Act of 1954, to transfer submarine nuclear technology to other states. He concluded that information on military power reactors could be exchanged but declined to offer an opinion on submarine reactors. Soon afterwards, Brownell sent a further letter to the Secretary of Defense, Charles Wilson, and the USAEC Chairman, Lewis Strauss, stating that the 1954 Act gave legal authority for the transfer of submarine reactor information but suggested that no action be taken without the informal approval of the JCAE.[9] Although Eisenhower agreed to postpone implementation of the 1956 Act, under Brownell's opinion he had legal justification to transfer nuclear submarine propulsion information. 'On February 5, 1957, he ordered the USAEC, DoD and the State Department to put into effect the July 1956 Agreement'.[10]

In May 1955, AERE's Head of Engineering Research & Development Division, B. L. Goodlet, visited the US where he had meetings with organisations interested in nuclear programmes. During discussions with Rickover, the Admiral asked Goodlet if the UK had a submarine project to which Goodlet replied that the UK did have a nuclear propulsion project. Rickover suggested that it would be a pity to cover the same ground and that it would be beneficial if the US and UK agreed to a joint nuclear propulsion programme. Rickover advised that if the UK had: '...a firm submarine project of our own he was certain that we would be given the information on *Nautilus* provided the approach was made through service channels and not through the A.E.C.'[11] Goodlet advised in his note to Harrison-Smith that our service people should know of Rickover's remarks. Harrison-Smith subsequently wrote to the Engineer-in-Chief to inform him of Rickover's comments.[12] A few days later, the Controller wrote to Mountbatten to advise that he had approved a power plant of between 10,000–20,000SHP suitable for a submarine, he noted that both Harwell and the commercial firms: '...have stressed the importance of having a specified project to work on'.[13] This approval was given on the advice contained in an appreciation by the Naval Section at Harwell on

[9] John Simpson, *The Independent Nuclear State: The United States, Britain and the Military Atom* (London, Macmillan Press, 1983), p. 116.

[10] Botti, *The Long Wait: The Forging of the Anglo-American Nuclear Alliance 1945–1958*, p. 178.

[11] TNA, ADM 205/106, (8g) Note, Goodlet to Harrison-Smith, 1 June 1955.

[12] TNA, ADM 205/106, Letter, Harrison-Smith to Mason 25/405, 7 June 1955.

[13] TNA, ADM 205/106, Letter, Edwards to Mountbatten, 17 June 1955.

Nuclear Propulsion in the Royal Navy (T.S.D.402/55), which was forwarded to a sub-committee of the Ship Design Policy Committee 15 April 1955.[14] This would appear to be the first record of Rickover's offer to help the Admiralty and seems to have acted as a spur to formalise the programme by selecting a submarine as the first vessel to fit a pressurised water reactor and to state the plant's power parameters.

However, Rickover's enquiry opens intriguing questions concerning his visits to the UK during the mid-1950s. Research has uncovered evidence that Rickover made an earlier, unrecorded, visit to the UK in December 1954. Apart from the document which highlights this fact, there are no references to this particular visit in any of the files researched and no mention has been made of it in previous historical articles on the subject of Rickover.[15] In Gordon's memoir, he recorded in the chapter, 'My eleven days in England with Rickover', that its purpose was: '…to gather all the scientific and engineering information that the British Government had accumulated from their own extensive nuclear research programs'.[16] Gordon noted that he never saw nor heard about the report again. For their endeavours, Rickover gave the team an extra day in London to complete some Christmas shopping. Gordon stated that the original flight the team should have taken crashed at Glasgow, killing all the passengers. Given there are no "official" records of this Rickover visit and suspecting that Gordon may have confused the year, a quick search of plane crashes in December 1954 noted just one in the UK. A BOAC Stratocruiser was on a scheduled flight from London to New York when it crashed at Prestwick airport (Glasgow) on Christmas Day 1954. Goodlet's meeting with Rickover came just five months after Rickover's initial visit to Harwell; it would be easy to imply from Rickover's question, and from Goodlet's position as Head of Research and Development at Harwell, that the 1954 visit had nothing to do with submarine propulsion. However, given the purpose of the visit, as described by Gordon above, and Rickover's sole determination to master every aspect of nuclear engineering for submarine propulsion purposes, this may be a naïve view. As discussed earlier with Rickover's involvement with Shippingport, his question to a British metallurgist (Finniston) in April 1955 would seem to suggest that British metallurgists were very well respected by Rickover and that

[14] DNP 2, NP184/2011, Board Direction for Prototype Submarine Design, p. 15.
[15] Gordon, 'Working for Admiral Rickover'.
[16] Gordon, 'Working for Admiral Rickover', p. 14.

they had information that was useful to him and the programmes he was involved with. A query to the United States Naval Historical Foundation received the reply: 'As [with] most clandestine visits in history, there are usually quid pro quos'.[17] Could it be that the quid pro quo in this case was Rickover's offer (through Goodlet in May 1955) to the UK to purchase a naval propulsion unit from the US? Without access to further documentation relating to Rickover's 1954 visit, it is doubtful we will ever find out if this was the case. Obviously, Rickover and his team had a very lucky escape; goodness knows what impact Rickover's death at that time would have had on the US nuclear propulsion programme or, indeed, the Admiralty's own propulsion programme. Despite the intrigue however, this visit falls outside the confines of this book and is detailed here merely to set straight the historical record of Rickover's visits to the UK.

What Price Exchange of Information?

Following discussions with Rickover during his 1956 visit, the E-in-C, Vice Admiral F. T. Mason, wrote a precis of Rickover's comments and views on the exchange agreement. Mason was keen that a study be made of all the implications of the exchange agreement on the freedom to build a nuclear merchant marine fleet unfettered by possible American contractual restraints.[18] There is evidence of uneasiness at the Admiralty and the UKAEA of accepting US nuclear information carte blanche. Sir Christopher Hinton pointed out that, with the UKAEA handling the transfer of submarine information, care had to be taken regarding their responsibilities not only to the submarine project but also to the development of nuclear propulsion for the merchant marine.[19] In an undated paper on the exchange of nuclear information, the author wrote that we are now too far committed on the present reactor to make much use of American information: '…though it might save some time spent investigating blind alleys'.[20] The Controller, Admiral Sir Peter Reid, was also wary of being restricted by the exchange of information and advised that it was important to

[17] Private correspondence, Commander D. Winkler USN Reserve, 29 March 2016.

[18] TNA, ADM 205/112, Office of the First Sea Lord, 1955–56: (10d) role of the submarine in peace and limited war, Note, Edwards to Mountbatten, 6 September 1956, paragraph 4g.

[19] TNA, AB 16/3162, Note, A.E.X.(57) 4th Meeting, 21 February 1957.

[20] DNP 1, ALO 02617/97 – 000378, Paper, Nuclear Propulsion Exchange of Information between the U.S. and U.K., 1957.

discriminate in the information the Admiralty obtained from the US.[21] Sir John Lang later confirmed this view noting that since the amendment allowing exchange of information in mid-1956: 'British work on marine nuclear propulsion has advanced, and the exchange of information now made possible is less valuable to us'.[22] Lang echoed Reid's view that the Admiralty was not presented with a mass of information which could have been useful two years previously, but which now might only serve to embarrass the Admiralty.

Lang proposed to discuss with the Americans and settle on a procedure whereby the Admiralty could select what nuclear information they required and decline information they already had. This stance, in the upper echelons of the Admiralty seems to be in total contrast with Macmillan's policy of nuclear interdependence. It appears that the Admiralty, although keen to accept some nuclear information to allow them to proceed on their submarine project with confidence, was equally keen to keep their reactor design independent of American control by limiting the information they were willing to accept and so be able, as the authority responsible to the marine industry in general, to proceed with nuclear propulsion for the merchant marine if such a programme was initiated by government. The Admiralty felt that the UKAEA should avoid any embarrassment to its civil marine programme which might be caused by accepting information from the Americans in connection to the submarine and might shackle British freedom of action developing nuclear propulsion for merchant ships. It was the Admiralty's view that the UKAEA should specify the information it required rather than leave it to the Americans to send what they think it ought to have. Strath was concerned that due to the delay since the Agreement was signed, the Americans may feel obliged to inundate the UKAEA with information.[23] A couple of days later, in response to the USAEC requesting information about Calder Hall, Lang informed Strath that he had been prepared to advise Sandys to abandon any agreement over the submarine reactor if it involved conditions which may embarrass us over marine nuclear propulsion. Lang also noted that the Admiralty was anxious to avoid being given information it already had, and that broadly speaking the Admiralty thought it had little to learn from the US as regards

[21] DNP 1, ALO 02617/97 – 000378, Letter, Controller to First Lord, 13 February 1957.
[22] TNA, AB 16/3162, Letter, Lang to Millar, 21 February 1957.
[23] TNA, AB 16/3162, Letter, Strath to Peirson, 6 March 1957.

the reactor for its first submarine.[24] I would suggest that this stance by the Admiralty does indicate a certain degree of confidence in the design of its own reactor.

Whilst Lang's statement may be true for the reactor physics, there is consensus among Royal Navy engineering officers from the period that without the adoption of US project management methods and the discipline of adherence to engineering standards and quality control, introduced by Rickover with the exchange of information, the UK would have struggled with the timescale of its own project. A typical observation from that cadre of engineering officer is summed up by Captain Colin Farley-Sutton, a member of the Dreadnought Project Team from 1960 and an Admiralty Engineer Overseer at Dounreay from 1964, who wrote that the project was saved a lot of time and grief because of the adoption of Rickover's management ethos: '…and the know-how derived from the US having "done it" many times before'.[25]

A few days later, Lang followed up his letter with a telegram in which he confirmed that if the UKAEA was pressed to envisage a package deal to exchange information in respect of civil power stations and the USS *Nautilus* installation, Strath need not be influenced by the threat of non-cooperation on the submarine reactor. Lang envisaged that Strath could use the reactor as a bargaining chip but wanted to keep the machinery separate as the Admiralty was keen to learn about the air-conditioning plant and other machinery which might be exchangeable under the military rather than civil Agreement.[26] The threat of non-cooperation did not occur as the USAEC wrote to the UKAEA to inform them that they wanted to proceed independently with the exchange of submarine information. A letter was forwarded advising what information the UKAEA would provide to the USAEC and what information the USAEC would provide in return. This included general information on the design, construction, operation and maintenance of submarine reactors and detailed information, within the above categories, on the USS *Nautilus* plant (including the prototype) and the S5W.[27] By separating the submarine reactor information from the civil nuclear information, the course was clear for the UKAEA to continue research on merchant marine propulsion,

[24] DNP 1, ALO 02617/97 – 000378, Letter, Lang to Strath, 9 March 1957.
[25] Private correspondence, Captain Colin Farley-Sutton RN Ret'd, 28 June 2018.
[26] DNP 1, ALO 02617/97 – 000378, Telegram, Lang to Strath, 12 March 1957.
[27] TNA, AB 16/3162, Letter, Willis to Hall, 17 April 1957.

should it arise. Conversely, the Admiralty would receive information which would enable it to compare its design to US reactors and discover exactly where they stood in relation to reactor development with the US.

Lang was Secretary of the Navy 1947–61, his biography describes him as: '…a formidable advocate for any policy which he decided to support'.[28] Much of his correspondence would appear to confirm his cautious nature, wanting to ensure every aspect of the problem is known prior to reaching a conclusive decision; Lang could be viewed as the consummate civil servant. However, Lang could be construed as having a negative attitude towards US assistance. In the correspondence reviewed, Lang merely defends the status quo vis-à-vis the UKAEA's position as sole manufacturer of fuel elements and of the proposed merchant nuclear marine project. Lang certainly does not come across as a keen supporter of the Royal Navy's ambition to acquire nuclear propulsion in a manner that can be seen in Mountbatten's and other senior naval officers' correspondence on the subject.

With the exchange of information formalised, Rear Admiral Wilson prepared for a meeting with Rickover to discuss wider issues, including the scope of information to be exchanged. Considering many in the Admiralty felt confident that the UK design was far enough advanced that there would be little, if indeed any, need for US information, a briefing note advised the possibility that Rickover may try to persuade the Admiralty to buy a US reactor.[29] It is not clear from the material researched why the prospect of purchasing a reactor from the US was brought up. Rickover's brief comments to Goodlet in May 1955 concerning the USS *Nautilus* were obviously known to the Admiralty but no other overtures are recorded. There were numerous questions being directed at USAEC staff which concerned Rickover causing him to snap at a meeting that detailed questioning was interfering with US reactor work. Rickover suggested that the Admiralty and UKAEA should send a technical delegation to the US to learn from the US programme to enable the UK to then decide its requirements.[30] It is possible that the Admiralty was aware that their questions were not focussed enough and possibly having a detrimental effect

[28] ODNB, Lang, Sir John Gerald (1896–1984). http://www.oxforddnb.com/view/10.1093 [Accessed 29 June 2018]

[29] TNA, AB 16/3162, Note for Discussion with R.A.N.P. and Head of M., no date, paragraph E.

[30] TNA, DEFE 7/2055, Telegram, WEP 508, 9 April 1957.

on Rickover's staff who were not only dealing with the S3W/S4W reactors being fitted into the *Skate* class but were also dealing with the development of the S5W to fit into the USS *Skipjack*. The Polaris submarine building programme was also getting underway in early 1957 creating additional work for Rickover's department. The scenario was possibly offered up as an option Rickover could make to alleviate the work load of his staff, there is certainly evidence from the UKAEA representative in Washington that the list of questions appeared to be an attempt to cover all aspects of submarine pressurised water reactor plants.[31]

Despite many in the Admiralty having confidence in their reactor design, what is evident from the records is Mountbatten's interest in purchasing a complete reactor unit from the US for installation in HMS *Dreadnought*.[32] This interest possibly indicates Mountbatten's frustration at the slow rate of UK progress (a reflection of Lang's cool reception towards nuclear propulsion?), or even his eagerness to get a Royal Navy nuclear-powered submarine to sea in the shortest possible timescale. Given his relationship with the Chief of Naval Operations, Admiral Arleigh Burke USN, Mountbatten was able to call on him for all manners of support; from assistance in nuclear training resources for the Royal Navy to briefing the UK Minister of Defence on the role of NATO navies in a global war.[33] If Mountbatten was to make a case for adopting the Polaris missile system to Cabinet Ministers and secure a future for the Royal Navy, then the successful and safe implementation of submarine nuclear propulsion technology was a priority of the highest order. Whether the power plant was of British or American origin appears to have been a secondary concern to Mountbatten.

The Offers to Purchase a US Submarine Reactor

Under the aegis of the 1956 Agreement, Eisenhower ordered the implementation of the exchange of nuclear submarine information in February 1957. The UKAEA was quick to begin organising a Technical Mission to the US comprising all elements involved in the submarine project.[34] However, before the visit could be fully organised, Mountbatten was

[31] TNA, AB 16/3162, Letter, Willis to Peirson, 4 April 1957.
[32] TNA, AB 16/3162, Letter, Strath to Peirson, 6 March 1957.
[33] Broadlands Archive, MB1/I391, Letter, Mountbatten to Elkins, 21 January 1957.
[34] TNA, AB 16/3162, Letter, Peirson to Strath, 19 February 1957.

strongly urged by Vice Admiral R. F. Elkins to invite Rickover to the UK: '...His visit to you last year made a deep impression on him and he still talks of it'.[35] Rickover accepted an invitation from Mountbatten and the visit took place 20–22 May 1957. The Technical Mission to the US was postponed until Rickover returned to the US so that a revised programme for the Technical Mission could be arranged resulting from discussions during Rickover's visit.[36] Rickover's visit included a top-level meeting at the Admiralty with officials from the Admiralty, the UKAEA and industry; arrangements were made for Rickover to visit Calder Hall and the premises of the firms involved with the project.[37] For their part, the Admiralty was keen to learn about the organisation in the US for building the USS *Nautilus*; this was due to increasing criticism from the UKAEA of the Admiralty management of the project which had already led to the creation of a post of Rear Admiral Nuclear Propulsion (RANP).[38] Indeed, Mountbatten and Reid held a private discussion about their organisation with Rickover who was described as being particularly forthright in his denunciation of the Admiralty's organisation.[39]

As a result of the May meeting, Rickover cancelled the preparations made with Washington for the Technical Mission visit and re-arranged the tour to what he thought would be the UK's advantage. The Nuclear Propulsion Technical Mission took place 10–25 June, which would be deemed an outstanding success. At the first meeting Rickover stated that the exchange on the US side would be limited to the Idaho prototype reactor (STR1), the USS *Nautilus* (S2W) reactor and the USS *Skipjack* class reactor (S5W). '...detailed information would not be passed on SSN 578 [*Skate*] class, *Seawolf*, or any General Electric work'.[40] There followed a series of lectures on the reactors outlined by Rickover; it is understandable that no information would be given on the USS *Seawolf*, which had a sodium-cooled (S2G, General Electric built) reactor; however, there appears to be no record of why the *Skate* class reactor was omitted from the exchange of information. There were four submarines in the class and

[35] TNA, AB 16/3162, Telegram, ABJSM to Admiralty, 11 April 1957.

[36] TNA, DEFE 7/2055, Telegram, BJSM to Admiralty, 17 April 1957.

[37] Broadlands Archive, MB1/N104, Memorandum, Rickover visit, 8 May 1957.

[38] TNA, AB 6/1867, Nuclear submarine (Dreadnought) project: organisation, 1956–58, Memorandum, Plowden, 13 March 1957.

[39] Broadlands Archive, MB1/I300, Quarterly Newsletter, 1 August 1957, p. 12.

[40] TNA, AB 16/2032, Nuclear Submarine Committee, 1957, Report, Visit of nuclear propulsion technical mission, June 1957.

due to Rickover's team being unable to reach a consensus on the best machinery plant layout, two plants were built, S3W and S4W; the only difference between the plants was the design of the Steam Generator and the reactor compartment layout.[41] The S3W plant was fitted in the USS *Skate*, the USS *Sargo* and the SSGN, USS *Halibut*; the S4W plant was fitted in the USS *Swordfish* and USS *Seadragon*.[42] The only possible reason for omitting information on the S3W/S4W plant was that the US Navy was already planning on taking advantage of experience gained from the USS *Nautilus* (S2W plant) which had shown small plants, such as S3W/S4W, were inadequate for high speed. During the visit, a trip was made to Groton to see the USS *Skate* fitting out and the USS *Skipjack*, which was under construction; it is evident the Americans did not exclude the British mission from visiting the USS *Skate*. With no information exchanged on the S3W reactor plant it is curious that Rickover would later propose that the UK purchase this plant. However, the first serious proposal to the UK to purchase a US naval reactor was made 5 October 1957 when Rickover offered a USS *Nautilus* plant (to the First Lord, Selkirk) at a cost between $11.4 M and $15 M.[43]

There does not appear to have been any Ministerial debate on the subject, which seems to highlight continuing political indifference to the programme; however, it is recorded by Sir Frederick Brundrett, the Royal Navy's Scientific Advisor, that Eisenhower and Macmillan had mentioned the nuclear submarine in conversation.[44] In mid-October, Mountbatten had arranged for Sandys and other high-ranking officials to go to sea in the USS *Nautilus* as a physical demonstration of a nuclear submarine's capabilities; it is entirely possible that the subject was brought up during the voyage. The merits of the proposal were, however, discussed by senior officials within the Admiralty and the MoD. Writing to Sir Richard Powell, the Permanent Secretary at the MoD, Lang outlined the Admiralty's views on Rickover's proposals advising that the acquisition of a reactor and the propulsion machinery to go with it would be a better proposition than the purchase of the reactor on its own.[45] Sir John Carroll, Deputy Controller of the Navy, was concerned at the thought of buying a "Chinese" copy

[41] Hewlett and Duncan, *Nuclear Navy*, p. 280.

[42] Hewlett and Duncan, *Nuclear Navy*, Appendix 2, p. 399.

[43] TNA, PREM 11/2554, Telegram, No. 2138, 19 October 1957.

[44] TNA, ADM 205/178, Office of the First Sea Lord, 1957–58: nuclear propulsion, Letter, FB/442/57, 20 November 1957.

[45] TNA, DEFE 7/2055, Letter, Lang to Powell, 22 October 1957.

and posed the question; what use could be made by Rolls-Royce if they were to buy replicas of the USS *Nautilus* from the US?[46] Mountbatten also discussed the situation and was advised by Wilson that if the Admiralty was to buy a complete propulsion unit then it was imperative that the Americans agreed to the Admiralty negotiating full commercial licences to enable the UK to manufacture its own future plants.[47]

Rickover advised in June that he wanted no more technical visits from the UK, his team was already busy with the *Skipjack* class of submarines and the new Polaris submarines, also to be powered by the S5W reactor. Rickover paid another visit to the Admiralty 24 January 1958, and during the meeting he emphasised how much money and effort had been spent in the US in pursuit of nuclear propulsion. Despite information being exchanged on the *Skipjack* (S5W) reactor since June 1957, Rickover offered the UK a chance to buy a whole *Skate* (S3W) class propulsion plant, from the reactor to the propulsion machinery. Rickover and Mountbatten had discussed the possibility of purchasing the whole machinery set prior to the main meeting and is documented as presenting the meeting with a "fait accompli".[48] It is recorded that Mountbatten was interested in buying a complete machinery unit as early as 1956; it is therefore highly probable that he and Rickover discussed the merits of purchasing a complete machinery set prior to the meeting.[49] Rickover stated that this would give the Admiralty practical knowledge and experience to facilitate their research in the field so when it came to building further submarines: '…we could stand on our own feet. He did not want to see us abandon our project'.[50] This is in contrast to Eric Grove's assertion that there was US pressure on the Admiralty to abandon the project; in making the offer the US wanted to delay the arrival of a competing design and influence British ideas in their direction: 'thus heading off an original competitive propulsion concept'.[51] It is a fact that the Dounreay prototype was designed to produce 20,000SHP whereas the S5W plant produced only 15,000SHP, as William Crowe noted: 'The Navy's scientists at Harwell were developing some original ideas which showed

[46] TNA, AB 16/1380, Letter, Cockcroft to Strath, 5 November 1957.
[47] TNA, ADM 205/178, Note, Mountbatten to Selkirk, 6 November 1957.
[48] Private papers, Rear Admiral Peter Hammersley.
[49] TNA, AB 16/3162, Letter, Strath to Peirson, 6 March 1957.
[50] TNA, DEFE 69/749, Minute, Selkirk to Sandys, 24 January 1958.
[51] Grove, *Vanguard to Trident*, p. 232.

considerable promise'.[52] However, the promise of something is one thing, but the successful delivery of that promise is another; there were, after all, technical and supply chain problems still being encountered.

Crowe was a Commander in the US Navy at the time of writing his thesis, and much of the data for Chap. 6, Atoms to Polaris came from three extensive interviews he held in the UK with individuals connected with the project since the early days. Crowe concedes that their testimony could not be confirmed in public documents, it is unfortunate that the individuals are not named in the bibliography to assist in locating the necessary documents in the National Archives.[53] However, from the extensive records researched, no evidence was found of the Americans attempting to coerce the Admiralty into abandoning their own research. Crowe notes that some opposition to the purchase of the S5W came from highly placed individuals rankled by Rickover's abrasive manner: '...the detached observer can hardly resist the conclusion that it was inspired by personal dislike rather than sober deliberation'.[54] There were high-ranking officers, such as Vice Admiral Elkins, British Joint Services Mission, Washington, who described Rickover as a "spoilt American child", but it is doubtful that they would have allowed personalities to interfere with their professional judgement.[55] Indeed, Elkins accompanied Rickover on his January 1958 visit to the Admiralty and had forewarned Brundrett that Rickover would suggest that they buy a reactor on favourable terms and: '...that we go on with our own development causing as little interference as possible with his'.[56] As seen in the previous section, there was indeed opposition to the purchase of the S5W and whilst some of this was due to loss of professional control, much of it revolved around the UK's independence to operate in the promising commercial aspects of marine nuclear propulsion that was foreseen in the mid-1950s.

Whilst it is tempting to correlate the Admiralty's reactor research with the UK's development of the hydrogen bomb which led Macmillan to

[52] William J. Crowe, 'The policy roots of the modern Royal Navy, 1946–1963', (unpublished PhD thesis, Princeton University, 1965), p. 261.

[53] Note: Admiral Crowe declined an invitation from Rickover to join the US nuclear submarine programme prior to studying at Princeton. He eventually became Chairman of the Joint Chiefs of Staff under President Reagan and was later appointed US Ambassador to London, 1994–97, by President Clinton.

[54] Crowe, 'The policy roots of the modern Royal Navy', p. 262.

[55] Broadlands Archive, MB1/I478, Letter, Elkins to Mountbatten, 24 September 1958.

[56] Hill, 'Admiral Hyman G. Rickover USN...', p. 9.

remark that 'in some respects we are as far ahead and even further advanced in the art than the Americans', the evidence researched does indicate that the Admiralty was having problems with the control rods and design of the fuel elements and would benefit from American information. It is of interest to note here that the detonation of the UK's hydrogen bomb in May 1957 led John Simpson to investigate the hypothesis that: '...the resultant Anglo-American agreements were negotiated from a position of British technological strength'.[57] Whilst this may have been true of the bomb makers, which, together with the launch of *Sputnik*, did allow for the eventual repeal of the McMahon Act and the signing of the Mutual Defence Agreement 1958, it is evident that this did not apply to nuclear propulsion. Article III, which allowed for the transfer of a submarine nuclear propulsion plant, also limited the exchange of information (and replacement cores and fuel elements) to a period of ten years.[58] As if to signify US dominance in the area of nuclear propulsion, the transfer of nuclear information and technologies in other areas (bomb making) continued. The decision to limit the nuclear propulsion part of the Agreement was political rather than military as it was Congress that legislated to allow the UK to purchase the S5W reactor.

Given Rickover's knowledge of UK research, it is difficult to conceive that he saw the Admiralty's research advanced enough to pose a threat to American dominance in the field. Indeed, one could question what the potential market was for nuclear submarine reactors, given that over sixty years later only six nations operate nuclear-powered submarines. The UK's market would have had little potential for growth with possibly only the Canadians and French buying from the UK. Furthermore, none of the private correspondence and papers from the officers interviewed hint at any evidence of US pressure on the Admiralty to abandon their research. Had such pressure been exerted, it is doubtful that the officers involved in the project would not have known of it. In discussions, the subject of US pressure to abandon Admiralty research was never brought up; all interviewees were adamant that the US offer was made on the basis to give the UK a datum from which to develop its own independent reactor research programme with a degree of confidence.[59] The final word on this subject

[57] Simpson, *The Independent Nuclear State*, p. xi.

[58] Mutual Defence Agreement, 1958, p. 3.

[59] Private correspondence and interviews conducted 2014 with Vice Admiral Sir Robert Hill, Rear Admiral Peter Hammersley, Captain John Jacobsen, Captain Colin Farley-Sutton, Commander Roger Berry.

has to be Rickover's; in a letter to the Commanding Officer of the USS *Nautilus*, Commander William Anderson, Rickover wrote: '...I did this because of my feeling of urgency about the international situation, my admiration for the British, and particularly my great liking for Admiral Mountbatten'.[60]

One part of the offer with the prospect to cause problems was Rickover's insistence that the proposal was based on a commercial contract between Westinghouse and Rolls-Royce. Essentially, Rolls-Royce was to manufacture the fuel elements under licence in the UK. Sir Edwin Plowden, Chairman of the UKAEA, wrote to Lang to record his opposition to the proposal as it was contrary to the Admiralty/UKAEA agreement on the submarine project, ran counter to government policy and was an uneconomical method of meeting the requirement.[61] Powell informed Macmillan that Rickover's offer would be discussed at the next Defence Committee meeting and that the possibility of Rolls-Royce manufacturing fuel elements needed to be resolved.[62] Subsequently, at the Defence Committee meeting, the offer was approved in principle and the First Lord was advised to urgently consult with the Chancellor, concerning finance, and the UKAEA, concerning fuel element production, prior to submitting the proposal to the Prime Minister.[63] Rickover's insistence on Rolls-Royce, however, removed the Admiralty's option to choose its own supplier. Up to this point Vickers Nuclear Engineering Limited coordinated the project work and Vickers-Armstrongs could have been expected to get the bulk of the contracted work on HMS *Dreadnought*. It would appear, that in reaching his decision to insist that the Admiralty employ Rolls-Royce as their commercial agents, Rickover was won over by the chairman of Rolls-Royce, Lord Hives, whom he met during a visit to the company in May 1957. Upon being introduced to Lord Hives, Rickover reportedly replied: 'A Lord eh? Chairman eh? And what are you then, a banker or a lawyer or what?' to which Hives replied: 'What me? No, no, no, no, no! Me, I'm

[60] Hill, 'Admiral Hyman G. Rickover USN...', p. 12.

[61] TNA, PREM 11/2635, Development of nuclear submarine: agreement on Anglo-US co-operation, 1957–59, Letter, CH(58)11, 4 February 1958.

[62] TNA, CAB 21/4852, Marine nuclear propulsion, 1957–61, Memorandum, RPP to PM, 5 February 1958.

[63] TNA, ADM 205/178, Minute, D.(58), 6 February 1958.

just a mechanic, just a bloody plumber'.[64] As Hill also noted, Rickover was not only lost for words, once Rickover realised that Hives had known Henry Royce, even Rickover could not dare to criticise the standards set by Henry Royce.

Choices and Decisions

With the offer on the table, there were now decisions to be made whether to accept the S3W plant or request the S5W plant. There was also an assessment to be made as to whether to purchase just the reactor plant and associated equipment, or to include the full propulsion machinery set. It has been asserted that: 'Rickover tried to fob off the British with the type of plant used in the U.S.S. *Skate*,' but this is untrue.[65] Rickover offered the *Skate* machinery plant because it was already a proven design; however, discussions began almost immediately on whether to accept the offer of the *Skate* plant or whether an attempt should be made to purchase the *Skipjack* plant which better suited the Navy's requirements.[66] Brundrett advised Sandys to accept Rickover's offer of the S3W plant in principle but to explore the possibility of substituting it for the larger S5W plant, the disadvantage being that USS *Skate* was a much smaller submarine, at 2900 tons submerged as opposed to HMS *Dreadnought*'s 4000 tons submerged.[67]

Mountbatten noted that Rickover's proposal would cost $22 M for the complete plant and the requisite training. As a comparison, US expenditure on nuclear propulsion research and development had been in the region of $750 M (approximately £268 M); the potential savings to the British project is evident.[68] Mountbatten also noted there was agreement that the preference was for the S5W plant; there was a further discussion with Rickover at which Sandys explained that the Royal Navy would prefer to have a submarine with characteristics akin to the USS *Skipjack*.[69] The advantages of a single screw and a "tear-drop" hull form, as demonstrated by the conventionally powered submarine USS *Albacore*, had been

[64] Private papers, Robert Hill, Lecture [notes] delivered at HMS Sultan 11 October 2012, p. 20.

[65] Philip Ziegler, *Mountbatten: The Official Biography* (London, Collins, 1985), p. 559.

[66] TNA, ADM 205/178, Memorandum, Mountbatten, 29 January 1958.

[67] TNA, DEFE 7/2055, Minute, FB/59/58, 28 January 1958.

[68] TNA, ADM 205/178, Memorandum, Mountbatten, 29 January 1958.

[69] TNA, DEFE 7/2055, Minute, FB/64/58, 30 January 1958.

recorded by the Royal Navy as early as February 1955.[70] At a submarine policy conference in February 1956, it had been recommended that the first nuclear submarine should have an *Albacore* hull. The Staff Requirement for HMS *Dreadnought* was issued 31 October 1957, and had similar characteristics to USS *Skipjack*.[71] In fact, the contracts for HMS *Dreadnought*'s hull and machinery had been placed with Vickers-Armstrongs between February and May 1957, which indicates how far the Admiralty's project had advanced.[72] Accepting the *Skate* machinery set would have involved a major re-drawing of the submarine's hull to accommodate two shafts, the difference in machinery weights and to take account of the lower power output of the S3W reactor plant; the S3W plant was rated 6600SHP whereas *Dreadnought* was designed for 15,000–20,00SHP (the S5W plant was rated 15,000SHP). Furthermore, the Admiralty's reactor had been designed along similar lines to the S5W plant so there could be little engineering objection concerning doubts that the UK-designed machinery could be matched to the S5W plant.[73] Rickover did not mind which power plant the Admiralty sought from the US and appreciated that the S5W plant was a more attractive proposition. It was advised that the S5W plant could cost as much as $29 M and take nine months longer to deliver; however, Rickover did advise that the decision over which plant to pursue was one for the Admiralty to make, he preferred the S3W plant as it was a proven design.[74] It is apparent from files researched that Rickover was happy to support the Admiralty's request to purchase whichever reactor the Admiralty decided upon without further influence from him.

With Rickover's tacit support secured, the Admiralty conducted an initial appraisal of the S3W and S5W reactor plants. Almost immediately the S3W plant was discounted as it was operational in a type of submarine which the Navy did not require, which, incidentally, was also the opinion of the US Navy which had cut the order of the *Skate* class due to their insufficient speed and capacity to stock the necessary operational equipment. After a short discussion it was concluded that due to the differences in the weight of the S3W machinery, the S5W reactor plant would be the

[70] TNA, ADM 205/106, (8b) Item 3, Future Construction, 7 March 1955.

[71] TNA, ADM 1/26843, HM Submarine *Dreadnought*: staff requirements, 1957–58.

[72] DNP 2, NP184/2011, Summary of Progress 1957, p. 23.

[73] TNA, DEFE 72/45, Relations with the USA stemming from the Bi-Lateral Agreement [...], 1954–68, Draft Appreciation of Admiral Rickover's offers, 31 January 1958.

[74] TNA, DEFE 19/50, Chief Scientific Adviser. Nuclear Propulsion, 1957–59, Note, Pritchard, 3 February 1958.

better proposition. The memorandum concluded that some thought be given to the different views before a final agreement was sought with the US.[75] A table produced by the Director Naval Construction Department illustrated the basic characteristics of a US/UK *Skate*, a US/UK *Skipjack* and *Dreadnought*. The differences were striking: endurance at a high speed of 2000 hours for a US/UK *Skate* compared to 3000 hours for a US/UK *Skipjack*, a thirty-day and a ninety-day patrol period, respectively, and a capacity to carry eighteen torpedoes against twenty-four torpedoes.[76] There were two further factors in favour of the S5W; it was designed for a single shaft, which would fit into HMS *Dreadnought* with little modification to the hull. Secondly, the Royal Navy's advanced 2001 sonar, under development at Portland, could be installed rather than a sonar with an inferior capability and range acquisition. Reid wrote that the S5W would enable the Admiralty to develop a nuclear submarine similar to the USS *Skipjack*'s performance and in some respects surpass it. Reid also noted that it could possibly lead to development: '…at a later date, of a British submarine armed with the Polaris weapon'.[77]

The other subject requiring a decision was to whether to purchase the reactor plant and associated control equipment only, or to include all the propulsion machinery. The merits of procuring the reactor and equipment within the reactor compartment, and of procuring all feasible US equipment aft of the forward reactor compartment bulkhead were subject to much debate. Mountbatten was in favour of purchasing the full propulsion set, as First Sea Lord he could use his influence but ultimately would have to regard the advice of the Controller (who in turn would be advised by E-in-C and other engineering department heads). Initially, members from the DPT, DNC and DEE departments concluded that it would be best if the US was requested to supply the reactor plant only, their reasons included; differences in machinery, British design effort to date would be wasted and a large number of conventional machinery orders would need cancelling. However, RANP thought it unwise to fit British conventional machinery without first testing it shoreside.[78] During a discussion on the merits of the *Skate* or *Skipjack* reactors, Rear Admiral A. G. Mumma USN,

[75] TNA, DEFE 72/45, Draft Memorandum, Appreciation of Admiral Rickover's offers, 31 January 1958.

[76] TNA, ADM 205/178, Table, D.N.C. Department, February 1958.

[77] TNA, ADM 116/6411, Nuclear propulsion unit for HMS/m Dreadnought, 1958, Memorandum, B.1191, February 1958.

[78] TNA, DEFE 72/45, Paper, 31 January 1958.

Chief of the US Navy Bureau of Ships, was reported as being in favour of the Admiralty buying *Skipjack* but: '…also expressed a strong opinion that we should obtain an entire plant from reactor to propeller'.[79] It is noted in the same correspondence that the Staff Engineering Officer (Propulsion) in Washington was against this advice; it would certainly seem at this juncture that the engineering advice was going against Mountbatten's plans. As late as May, Mountbatten was advised that it would be helpful to inform Mumma and Rickover that the Admiralty would probably order the full machinery set, which indicates that a decision had still to be made.[80]

The Chancellor, Derek Heathcoat-Amory, replied to Macmillan's request for his views and advised that he was prepared to accept the offer of the *Skate* plant on condition that the overall cost would be about £8 M, that expenditure over the next three years would not exceed what was already planned and that dollar expenditure was kept to a minimum. Heathcoat-Amory also advised that Rolls-Royce should not have a monopoly on the fuel elements.[81] The Chancellor also wrote in reply to Sandys the same day; there is no mention of the USS *Skipjack* alternative in either of the Chancellor's replies and it has to be assumed that he gave his views only on the cheaper, S3W plant offer.[82] It is clear at this stage, when a decision upon which plant to adopt had yet to be made, that the Admiralty was pushing hard on the Treasury for the S5W plant.[83] A decision had still to be made when the Ambassador in Washington, Sir Harold Caccia, wrote to the Secretary of State, John Dulles, to advise that he had been instructed by HM Government to enquire whether the US Government was prepared to sell a complete nuclear propulsion plant as laid out in the enclosed Admiralty's memorandum.[84] After much correspondence between officials and Ministers stating the military advantages of the S5W plant and corresponding savings to be made in research and development, the Chancellor wrote to Sandys to agree to the decision to buy the *Skipjack* machinery at a cost in the region of $23 M (approximately

[79] TNA, ADM 116/6411, Letter, Elkins to Mountbatten, April 1958. Note: Rear Admiral Mumma was Rickover's Naval Head of Department.

[80] TNA, DEFE 72/45, Letter, Elkins to Mountbatten, 16 May 1958. Note: Vice Admiral R. F. Elkins was Admiral British Joint Services Mission (AJBSM) in Washington.

[81] TNA, PREM 11/2635, Note, Chancellor to PM, 5 February 1958.

[82] TNA, DEFE 13/523, Minister's Office: American offer of submarine nuclear propulsion unit, 1958, Note, Chancellor to Sandys, 5 February 1958.

[83] TNA, ADM 116/6411, Note, RANP 15/2/58, 14 February 1958.

[84] TNA, DEFE 13/523, Letter, Caccia to Dulles, 3 March 1958.

£9 M) and that the First Lord was to notify the US Government accordingly.[85] In May, the Technical Chief Executive, Baker, chaired a meeting to discuss the purchase of S5W and main propulsion machinery and decide on the extent of the machinery to be purchased.[86] The US was duly informed that the whole *Skipjack* machinery set was to be purchased with the exception of the propeller, shaft and other itemised equipment. In June, the USAEC agreed to an amendment to the Atomic Energy Act, 1954; in effect this would allow the acquisition of the S5W plant by the UK Government as enacted in the Mutual Defence Agreement, 1958.[87] With political acceptance of the Royal Navy's requirement for the S5W plant, and commercial discussions proceeding between Westinghouse and Rolls-Royce, the Royal Navy could now begin to see the dividends of its formation of the Dreadnought Project Team and management of the project.

These events suggest that Mountbatten did meet with resistance to his plan to purchase the whole machinery set rather than attempt to adapt prototype British propulsion machinery to the S5W plant, which may have been met with difficulties. In later life, Mountbatten seems to have been uncertain if the correct decision had been made; when Mountbatten met Jacobsen at Chatham Dockyard in 1977, he asked if he had done the right thing to which Jacobsen replied he had.[88] Two years later, visiting the Royal Naval Engineering College at Manadon, he posed the same question to Hammersley, noting that he had gone against the advice of his top engineers; Hammersley also replied that he had made the correct decision.[89] However, one "top engineer" had been in favour, Baker, who, it was reported, deemed that the Dreadnought project had been a success in part due to Rickover selling Britain the S5W plant and: '...second to me for insisting that this S5W be used by us in an environment similar to *Skipjack*'.[90] Baker had the final say in the Dreadnought Project Team, and it is assumed that Mountbatten took the advice of the meeting Baker had chaired and disregarded the dissenters who were possibly looking after the interests of their own departments. They had, after all, initially objected to

[85] TNA, DEFE 7/2055, Minute, D-S 535, 21 April 1958.
[86] TNA, DEFE 69/749, Minutes, Baker, 21 May 1958.
[87] TNA, PREM 11/2635, Telegram, No. 1423, 7 June 1958.
[88] Private papers, Captain John Jacobsen RN Ret'd.
[89] Private papers, Rear Admiral Peter Hammersley.
[90] David K. Brown, 'Sir Rowland Baker, KB, RCNC: A personal appreciation', *Journal of Naval Engineering*, Vol. 36, Book 2, (Dec., 1996), 327–48 (p. 343).

attempts to set up the combined engineering project team that became the DPT in late 1957.

Mountbatten Corrections

During research it became apparent that there are anomalies of events in Mountbatten's official biography by Philip Ziegler. By his own admission, Mountbatten had very little information in his archives relating to the development of nuclear propulsion. It was Mountbatten's intention to write the events from memory and then correct them from records held at the National Archives and with input from people who had the experience to contribute to, and indeed, correct, his account. Mountbatten forwarded a rough draft of the relevant chapter to Hammersley for him to comment on; Mountbatten advised: 'There is no hurry about a reply as I, myself, am off to Ireland for a month at the end of this week'.[91] Mountbatten was murdered by the IRA 27 August 1979; his draft to Hammersley went uncorrected and parts are in his official biography. It is recorded that Hammersley thought long and hard about what to do about the letter and its draft enclosure before deciding to do nothing. Hammersley regretted his decision as the official biography contains some errors which Mountbatten: '...would not have known about, and I wish that I had corrected them'.[92] Ziegler's biography references the draft copy in footnote 40 on page 557 (MB1/K208A) and although not all the draft was used in the biography, the errors contained in the draft and the biography are addressed here for future reference (Fig. 5.1).

Mountbatten took up his appointment as First Sea Lord 18 April 1955, at which time there had still been no formal decision as to which type of vessel the first PWR nuclear propulsion reactor should be fitted with. As discussed in the sub-chapter, "Admiral Hyman G. Rickover USN", an appreciation by the Naval Section had advised that a submarine should be the first vessel fitted with a pressurised water reactor and had already been submitted to a sub-committee of the Ship Design Policy Committee a few days prior to Mountbatten's appointment. In July, the Director of Plans advised Mountbatten that a proposal would shortly be placed before the Board to design and build a nuclear submarine, money for the submarine

[91] Private correspondence, Mountbatten to Hammersley 31 July 1979.

[92] Private papers, Rear Admiral Hammersley speech to IMarEST lunch club, 17 September 2004.

Fig. 5.1 Rear Admiral Peter Hammersley RN, circa 1979

had been included in the long-term defence programme.[93] As examined in Chap. 3, the Admiralty's letter to the Treasury outlining the project was sent 8 December 1955 and formal authority to proceed was received 6 January 1956. Mountbatten's notes and Ziegler's biography refer to a meeting at which Mountbatten arranged for a model of a nuclear submarine to be commissioned and placed opposite the Chancellor's seat, whom they identify as Heathcoat-Amory. Supposedly, this was the meeting that Mountbatten describes as getting our: '...first nuclear submarine authorised'.[94] Heathcoat-Amory was Chancellor January 1958 to July 1960; the Chancellor in 1955 was R. A. Butler with Harold Macmillan taking over the office from December 1955 to January 1957, it is entirely possible that Mountbatten was referring to a Meeting discussing the decision to purchase the S3W or S5W reactor plant. As to whether the holder

[93] TNA, ADM 205/106, Letter, Coke to Mountbatten, 19 July 1955.
[94] Ziegler, *Mountbatten*, p. 557.

of the second most powerful office in Government could be swayed in a momentous financial decision by the trinket of a model submarine, even by someone of Mountbatten's stature, is open to debate. The only model referred to in the files researched is of the *Skipjack*, which was offered to Mountbatten by Rickover in 1958; there is no record as to whether Mountbatten received the model.[95] There is, however, a model which matches Mountbatten's description in the Office of Naval Reactors, Washington D.C.[96]

As recorded earlier, it is documented that Mountbatten and Rickover held discussions on the merits of procuring not only the reactor plant but also all feasible US machinery aft of the forward reactor compartment bulkhead. According to Mountbatten's version of events he presented his case for purchasing the whole machinery set to the Committee at the main meeting. On completion of the presentations, he advised that he would report to the First Lord and get a Board of Admiralty decision on this important matter due to the considerable disagreement among the Committee. 'I was successful in getting the complete SKIPJACK propulsion plant including the steam turbine from the Americans included in the DREADNOUGHT'.[97] It has been presented in the previous chapters from archival evidence that the offer at the 24 January meeting was for the *Skate* (S3W) plant and machinery set. No decision was made on exactly what to purchase for some time after until the Minister of Defence and the Chancellor both agreed to the Admiralty's case to purchase the *Skipjack* machinery and the Foreign Office was advised, some three months later, to inform the British Ambassador in Washington, Harold Caccia, accordingly.[98]

The idea that "Rickover tried to fob off the British with the S3W" has already been addressed in this chapter and evidence presented. One further anomaly remains in the biography causing confusion over the S3W and S5W plants. The story of Mountbatten seeing a newer version (the more powerful S5W reactor) in the *Skipjack* when he visited in October 1958 after which: '...there were no more suggestions that the British might make do with *Skate*', does not fit in with the narrative and the

[95] TNA, ADM 1/27198, Correspondence with Admiral British Joint Services Mission [...], 1958–59, Letter, Thistleton-Smith to Mountbatten, 14 November 1958.

[96] Discussion with Rear Admiral Steve Lloyd CBE, 20 January 2019.

[97] Private papers, Rear Admiral Peter Hammersley.

[98] TNA, DEFE 72/45, Minute, M.I.618/2/58, 30 April 1958.

evidence as presented.[99] Mountbatten is recorded as having visited the USS *Skipjack* during his US/Canada tour in October 1958 but this was too late to have any effect.[100] The decision to purchase the S5W plant had already been made and the MDA Act, allowing for the purchase of the S5W and complete machinery set had already come into law that summer.

In reference to future submarine propulsion units Mountbatten appears to have overlooked two basic facts: the UK's pressurised water reactor and main propulsion machinery were already well developed by 1958 and it was already approved that future submarines would have an all British reactor plant with Core A based upon the S5W core and a British-designed main machinery set. There was never any intention to make "Chinese" copies of the S5W plant, only to use it as a datum from which to develop the UK's own pressurised water reactor. With no further information on nuclear propulsion reactors forthcoming from the US, any UK-produced copies of the S5W plant would have soon become moribund and obsolete. Mountbatten wrote: '…I suggested that we should go on with the SKIPJACK equipment for the future'.[101] Mountbatten thought that there was a plot to fit British-designed propulsion plants as soon as he left the Admiralty; however, Mountbatten was wrong, this was never the case. Hammersley later noted: 'I do not believe that there was any question of Valiant and later Submarines being anything other than all British'.[102] Indeed, Admiral Sir Peter Reid, the Controller, wrote of the Admiralty's gap in reactor knowledge that, it is only in designing our own core that we gain the necessary understanding to diagnose and cure any defect which may arise in the reactor. 'Without this knowledge we are tied to American designed cores of a specific power; and if things go wrong with them, we cannot get them right without recourse to the Americans'.[103] The limitations of sticking with the US-designed S5W core is borne out by the official records at the National Archives and the evidence has been presented.

[99] Ziegler, *Mountbatten*, p. 559.

[100] Broadlands Archive, MB1/I478, Report of First Sea Lord's Visit to Canada and the U.S.A., para. 28.

[101] Private papers, Rear Admiral Peter Hammersley.

[102] Private correspondence, Rear Admiral Peter Hammersley to Vice Admiral Sir Robert Hill 10 February 2005.

[103] TNA, ADM 1/27750, Submarine machinery future research and development, 1960–61, Memorandum, B.1362, October 1960.

Management and Establishment of the Dreadnought Project Team

Mason's precis of his discussions with Rickover in 1956 also recorded Rickover's views on the management of the project. Rickover was emphatic that the organisation: '...required a man with "fire in his belly" [...] the remainder of us came under the category "quiet and contented cows"'.[104] Mason had already been thinking about changing the management structure but his discussions with Rickover altered his view on how it should be completed; the main recommendation proposed by Mason was the appointment of a deputy to oversee responsibility for nuclear propulsion. Mason recommended his deputy for Fleet Maintenance Rear Admiral G. A. M. Wilson for the post: 'I think he would fulfil Rickover's idea of having "fire in his belly"; certainly he is not a "quiet contented cow"'.[105] Wilson was appointed RANP in February 1957.[106] However, it was not only Rickover that was critical of the Admiralty's management of the project, the UKAEA was also concerned.

At a meeting of the Atomic Energy Executive (AEX), Lord Portal thought, with the possibility of information being exchanged on submarine reactors soon, it would be useful to review the progress and organisation of the project. Sir Christopher Hinton shared Sir John Cockcroft's view that a comprehensive examination of the organisation was required before detailed criticism and remedial proposals could be produced.[107] It was noted that the Admiralty and Vickers-Armstrongs were aware of some weakness in the project and it was thought that they would welcome an offer by the Authority to conduct an examination of the engineering aspect. Subsequently, Plowden visited Lord Weeks, Chairman of Vickers-Armstrongs, to express his uneasiness with the project. Weeks advised that he was confident with the organisation but would welcome advice from Risley on the engineering side. Plowden informed Weeks that if the project was to be a success, a good project engineer should have complete responsibility for management of the project. Plowden advised Weeks that he intended to have a senior engineer from Risley chair the Advisory Engineering Committee and suggested when he returned from the US in

[104] TNA, AB 6/1051, Letter, Edwards to Cockcroft, 5 September 1956, (enclosed Minute, 20 August 1956).

[105] TNA, ADM 205/112, (10d) Note, Edwards to Mountbatten, 6 September 1956.

[106] DNP 1, ALO 02617/97 – 000378, Memorandum, 36/57, 15 February 1957.

[107] TNA, AB 16/1379, Note, A.E.X.(57) 3rd Meeting, 7 February 1957.

March that he might want to hold discussions with Hinton on how Risley could help with the organisation.[108]

The head of the Physics Division at Harwell, J. V. Dunworth, had held a series of meetings with Harrison-Smith and one with Wilson, subsequently he wrote to Basil Schonland, Deputy Director of Harwell, with his findings. Dunworth noted that within the Admiralty organisation, Harrison-Smith and Wilson had no control over Naval Construction (DNC) nor the Instrument and Control Group (DEE). Dunworth also noted that the DNC and other Admiralty Departments would oppose any erosion of their authority: '...Harrison-Smith has been trying to do a very difficult job without adequate authority'.[109] Dunworth also reported that the Vickers Group had no proper executive control over its various business units and concluded that the Admiralty was not prepared to give authority over the project to one man and that the firms were top heavy. By mid-March, the AEX had received a report on the submarine project compiled by Hinton at the request of Plowden. Hinton cautioned that his conclusion: '...that the overall organisation is unsatisfactory', was based on information from Rolls-Royce and Admiral Rebbeck of Vickers-Armstrongs, no discussions were held with the Admiralty.[110]

Hinton describes the disparate Admiralty organisation thus: the hull of the submarine is being designed by DNC whose Department is based at Bath, the propulsion systems have to be manufactured to the requirements of E-in-C's Department also at Bath, although E-in-C has his office in London. RANP has recently been appointed and has an office in London, Harrison-Smith heads a small team at Harwell involved in experimental work which also considers and approves all proposals and designs for the reactor which are developed by Rolls-Royce in Derby. The design and construction of the power plant is the responsibility of the Vickers/Rolls-Royce Group comprised of Vickers-Armstrongs, Rolls-Royce and Metropolitan-Vickers (their offices based at Barrow, Derby and London, respectively). The shielding was being designed under the direction of Dr Forsythe who is based at Barrow; however, his team carrying out the work are located at Southampton. Hinton could not see a submarine being developed within a reasonable period without an integrated design team composed of members of the three firms and a senior member of the DNC

[108] TNA, AB 16/1379, Note, Plowden to Weeks, 13 February 1957.
[109] TNA, AB 6/1867, Memorandum, Dunworth to Schonland, 28 February 1957.
[110] TNA, AB 6/1867, Note, A.E.X.(57)17, 8 March 1957.

Department working in a single office. Hinton advised that this integrated team should be responsible to a single senior officer at the Admiralty. Wilson, who was considering the management of the project, noted in the report of his visit to see Hinton that the project was: '...being handled in far too many places without any real effective co-ordination'.[111] It can be appreciated how awkward it must have been to work between so many different groups at various geographic locations, with several managers responsible for the different aspects of the project, and it goes to the heart of Rickover's philosophy of responsibility and having one person answerable. It was a philosophy that took the Admiralty some time to adopt.

Plowden called to see Lang at the Admiralty and showed him Hinton's report but did not leave a copy. Lang appreciated Plowden's candour and wanted to discuss with Hinton and the Controller measures to adopt to improve the organisation.[112] Lang and Rear Admiral Peter Dawnay, the Deputy Controller, met with UKAEA representatives to discuss the criticisms and formulate plans to restructure the project organisation. The Admiralty members were advised that if the UKAEA was responsible for the project they would look to design the whole power plant from a single office as opposed to what was currently happening. Wilson too, was looking at means of changing the organisation but had only been in post a month and needed time to familiarise himself with the administration of the project and the responsibilities of the various component agencies. On his return from the Technical Mission visit to the US, Wilson reported to Reid that the US displayed a firm cohesive direction in its organisation which was: '...lacking in ours both on the Admiralty and Contractual side'.[113] A directive was issued by the Controller 16 August with the intention of strengthening the Admiralty organisation with the creation of a team of specialists advising on technical and production decisions under the authority of a team leader. The leader of the team would only have to seek higher authority if there were significant changes affecting costs or performance. A Dreadnought Task Group meeting was held 21 August to give effect to the directive and a draft Terms of Reference, C.E.4759/57, was circulated 28 August.[114]

[111] DNP 1, ALO 02617/97 – 000378, Letter, Wilson to Controller, 20 March 1957.
[112] TNA, AB 6/1867, Note, Plowden, 13 March 1957.
[113] DNP 2, NP184/2011, Organisation of the Project Team, p. 30.
[114] DNP 2, NP184/2011, Organisation of the Project Team, p. 32.

The Admiralty lost no time in moving its personnel and contractors to central locations, the same day C.E.4759/57 was circulated, Harrison-Smith wrote to Schonland to advise that as part of the reorganisation, personnel not involved with the operation of experimental equipment, such as Neptune and LIDO, would be removed from Harwell.[115] Harrison-Smith and others would move to Bath and Jack Edwards would remain as senior naval representative in charge of the remaining RNSS staff. Schonland was also advised that a combined contractors' team was being set up at Derby to design the reactor and its primary circuit.[116] Roger Berry was appointed to the Naval Section in January 1956 and recalls of his time there that Harwell: '…under the direction of Sir John Cockcroft, did not appear to be very interested in the Naval Programme'. Berry notes his own role at Harwell as maintaining contact with the E-in-C's Submarine Section, whose interests were the new *Porpoise* and *Oberon* class submarines, HTP submarines and the development of recycle diesel. '…they made it fairly clear that they did not want to get involved with nuclear,' Berry captures, from a junior member's perspective, the apparent disinterest of both the AERE and his contacts in the E-in-C's Department.[117] Berry describes Harrison-Smith as pleasant, but aloof, and being away frequently for long periods visiting Dounreay and Derby. Berry also commented that it was never made clear to him what his responsibilities were and that other members of the Naval Section seemed to be doing "his own thing". Berry's notes give an insight, at a local level, into the apparently muddled management of the Naval Section. The notes were written in 2009, and by his own admission Berry did not enjoy his time at Harwell noting that it had been a wasted opportunity due to the lack of direction which was only resolved with the formation of a unified project team. Berry was associated with the programme for many years having moved to Bath as a member of the Dreadnought Project Team, he was later appointed to the Dreadnought Liaison Engineers (DLE) team, under Captain Cotman, in the US and was based at Groton, Connecticut.

Now that the project organisation was being dealt with, Schonland wrote to Wilson to offer some thoughts on how the organisation should

[115] LIDO was an enriched uranium-fuelled "swimming pool"-type reactor (the core was suspended from a gantry in a concrete water tank) and was partly used for reactor shielding studies for the Royal Navy.

[116] TNA, AB 6/1737, Access Agreements: Admiralty submarine project; Rolls Royce, 1955–57, Memorandum, 280/96, 28 August 1957.

[117] Private correspondence, Commander Roger Berry RN Ret'd, 3 May 2013.

be managed and to seek clarification on the Authority's role in the project. Wilson replied that the Naval Board had approved the formation of a Dreadnought team which was being centred on the engineering departments at Bath, initially under the management of Harrison-Smith. This would enable a project leader to have direct contact with specialist electrical and engineering technical advice and have overall control for the project; the team being removed from Harwell would form the focus of a new reactor specialist section. Wilson draws attention to the formal agreement, C.P.60801/56, between the Admiralty and the UKAEA; this letter, dated 8 February 1957, was the document against which Wilson felt that co-operation should be measured.[118] In his reply Wilson also gave Schonland an outline of the Dreadnought organisation in which he reiterated that: '…the Project Team Leader will be the key man for the whole project,' however, Schonland disagreed; as he viewed the new organisation, it was the Deputy Engineer in Chief who was the real controller of the project.[119] It is apparent that the Admiralty and the UKAEA had differing ideas of what form the organisation should take and it could be argued that it was the Admiralty that was unwilling to adopt the business model required to make the project a success. Although not directed at officials in the Admiralty, Wilson's frustration at the lack of progress in the project is apparent when, in relation to waiting for Ministerial/Treasury approval to proceed with the purchase of highly enriched uranium, he commented: '…vacillation of this kind, which is symptomatic of the uncertainty surrounding this project, can only result ultimately in delay in completion'.[120]

The Dreadnought Propulsion Project Team (DPT) was formed under Office Memorandum 214/57, dated 8 November 1957. The memorandum acknowledged that the project required a novel approach and adaptation of the normal Admiralty organisation, the key responsibilities were outlined, and channels of communication highlighted. It was made clear that all departments involved were expected: '…to collaborate with the Team and to give full and prompt assistance whenever this is requested'.[121] The DPT comprised the following sections and responsibilities; DPT1—Dounreay Submarine Prototype, DPT2—Mechanical Engineering Ship Section (Secondary Systems), DPT3—Rector, DPT4—Electrical Section,

[118] TNA, AB 6/1867, Letter, RANP.48/9/57, 24 September 1957.
[119] TNA, AB 16/1380, Letter, Schonland to Strath, 1 November 1957.
[120] DNP 2, NP184/2011, Note, Wilson, 7 January 1958, p. 36.
[121] TNA, ADM 205/178, Memorandum, 214/57, 8 November 1957.

DPT5—Reactor Control Systems, DPT6—Primary Systems (Excluding Reactor) and DPT7—Nuclear Safety.[122] According to official papers, the DPT began work in its full integrated form on 18 November; however, there appears to have been a delay between its formation and it becoming fully functional.[123] From private papers and correspondence it is apparent that some early members of the DPT did not join at its formation and whilst this could be due to the appointments system in the Royal Navy, one member who joined in December 1957 gives a different view. Although the Board of Admiralty had instructed the Secretary of the Navy to set up the DPT at Bath with staff from three Departments, there was a delay, which has been described as a rear-guard action by Lang: '…who had refused to sanction the provision of office furniture etc for a project he deemed to be unnecessary'.[124] On reflection it has to be questioned why the Admiralty set up the DPT with a Captain as team leader, even on a temporary basis. Initially, Harrison-Smith was appointed team leader (possibly to provide continuity); he was soon to be relieved by Captain Terence Ridley. A Captain in the Royal Navy is a rank below Commodore, "a one star rank", and it could be strongly argued he would not have the authority to demand the resources required from the other Admiralty departments. Wilson was a Rear Admiral, "a two star rank", but his role within the organisation was subject to review.[125] Under Wilson's planned reorganisation it was he who controlled the team leader and the firms, the team leader had no control over the firms, only the Admiralty Departments.[126] It is evident from the files researched that the Admiralty thought the organisation as conceived would suffice to address the criticisms of the organisation's failings, it manifestly did not.

Although the organisational changes had yet to bed in, the Admiralty still received strong criticism from all participants, including naval officers. Sir Frederick Brundrett wrote to Mountbatten to inform him that he had received: '…very bitter criticism that the Admiralty organisation for handling this business is deplorable'.[127] The same day, Lang received a letter from Plowden strongly condemning the weakness of control and direction

[122] Private papers, Commander Roger Berry RN Ret'd, and Captain Colin Farley-Sutton RN Ret'd.

[123] TNA, ADM 205/178, Letter, Controller to Mountbatten, 19 November 1957.

[124] Private correspondence, Captain John Jacobsen RN Ret'd, 27 June 2018.

[125] TNA, ADM 205/178, Letter, Controller to Mountbatten, 19 November 1957.

[126] TNA, AB 16/1380, Letter, Schonland to Strath, 1 November 1957.

[127] TNA, ADM 205/178, Letter, FB/442/57, 20 November 1957.

given by the Admiralty.[128] Whilst some of the criticism was possibly due to delays in getting the DPT functioning properly, it was surely because the control of the organisation did not seem strong enough to manage the project and elicit the required assistance of the other Departments. To address Plowden's criticisms, Lang held discussions with Reid and the relevant Admiralty Departments after which he wrote to Mountbatten and Selkirk wanting an opportunity to explain the conclusions reached and their proposals. Essentially, the key feature of their proposals was: '…the concentration of executive control of the project in a Chief Executive who possesses the technical qualifications, status and driving force necessary for the discharge of the functions proposed for him'.[129] Lang and Reid subsequently held discussions with Strath in which they advised that it was the Admiralty's intention to appoint Rowland (Roly) Baker as Chief Executive. Baker was a Chief Constructor in the RCNC, a civilian rank equivalent to Rear Admiral, importantly, it was envisaged that the Chief Executive would have direct access to: '…the Director of Naval Construction, Engineer-in-Chief and Electrical Engineer-in-Chief in any case where there is a deadlock in his dealings with these Admiralty Divisions'.[130] A meeting was held the following day with Strath, members of the Admiralty, Rolls-Royce and Vickers-Armstrongs. Lang allowed the firms to take away the proposals and asked them to send their opinions. In early January, Baker, whose appointment was yet to be approved by the Treasury, was being introduced to the component parts of the organisation such as members of Vickers-Armstrongs and the Chief Engineer at Risley. Baker's formal appointment and the organisation for control of the DPT was circulated in Office Memorandum C.2/58; finally, the Admiralty had someone ultimately responsible for the whole nuclear submarine propulsion project.[131]

Purchase of the S5W Reactor

As with any piece of quality-controlled engineering there is a question of inspection and, crucially, competency. With the signing of the Mutual Defence Agreement 3 July 1958, the Admiralty was able to purchase the

[128] TNA, AB 6/1380, Letter, CH(57)98, 20 November 1957.
[129] TNA, ADM 205/178, Letter, Lang to First Lord/Mountbatten, 4 December 1957.
[130] TNA, AB 16/1380, Note, Strath, 19 December 1957.
[131] TNA, DEFE 69/749, Memorandum, C.2/58, 14 February 1958.

S5W reactor; however, questions soon surfaced surrounding inspection of the equipment to be supplied and the competency of the people doing the checks. During initial negotiations, the UKAEA stated that in order to be satisfied that the plant was safe they would require a competent US organisation to inspect the equipment and supply certification that it was to specification. However, the Admiralty drew the UKAEA's attention to Rickover's statements to the Joint Congressional Hearings on the amendment to the US Atomic Energy Act during which Rickover made it clear that he did not visualise the US Government having any responsibility with regard to the safety of UK-purchased equipment. In buying the reactor on a commercial basis, Westinghouse would make a nuclear core for the UK in the same factory as the American cores and the UK would provide inspectors to examine the various items and choose whether to accept them. 'When they accept delivery, if anything goes wrong it is their business. We will be out of it as a government [...] and have no further responsibility'.[132] The problem was that the S5W plant was being built to American standards, so the necessary documentation and data needed to be supplied for the equipment to be inspected against. In reply, the UKAEA advised that Rolls-Royce and Westinghouse should provide the necessary inspection procedure for the Admiralty to arrange independent inspection by either an organisation such as Lloyds or the Admiralty's own engineering surveyors.[133] Wilson advised Reid, Baker and others that accepting the machinery from the US was a UK responsibility and, therefore, it was important to urgently select the inspection staff and put them through a suitable course of nuclear training before they proceeded to the US.[134]

Baker headed a team to Washington to obtain information on the S5W plant to enable the design of HMS *Dreadnought* to be completed from a shipbuilding point of view. HMS *Dreadnought* would eventually be constructed of an entirely UK-designed front end encompassing a new bow array sonar, torpedo handling and discharge system and other unique British advances. The after end would require a complete redesign to accommodate the S5W reactor, the propulsion machinery and the auxiliary machinery as it was intended that the machinery arrangement would

[132] TNA, AB 8/791, Technical assessment: Dreadnought reactor, 1957–59, Letter, Peirson to Farmer, 18 July 1958.

[133] TNA, DEFE 69/749, Letter, Peirson to Wilson, 23 July 1958.

[134] TNA, DEFE 72/45, Letter, RANP 26/7/58, 24 July 1958.

be identical to the USS *Skipjack* layout.[135] During these discussions, Baker was advised that the Naval Reactors Branch would not make available the Safety Committee report on the reactor or any information on discussions between them and the USAEC Safety Committee. They would, however, provide Westinghouse with sufficient documentation: '...to enable the UK Safety Committee to make its own assessment'.[136] Although there would have been issues pertaining to intellectual property rights, design rights, and other commercial considerations, this decision was to be a handicap to UK designers. With no design philosophy to indicate how the present design had been achieved or what areas of research had been abandoned, all the UK had to go on was their research and the S5W plant as purchased. Without the USAEC Safety Committee report on the S5W plant, the Admiralty was getting the absolute basics to give a foundation to develop its own generic reactor design. They were certainly not getting the entire US knowledge and competence they would have liked to receive. Indeed, from the beginning of the Admiralty's nuclear submarine propulsion project, the US had refused to offer any critique of the Admiralty's own programme as this could have been interpreted as the UK benefitting from US nuclear knowledge (Fig. 5.2).[137]

With the purchase of S5W plant, it was the Admiralty's intention to continue development based on the current design of the *Dreadnought* machinery except for the core and control mechanisms: '...and any other items where advantage can be taken by adopting the American design'.[138] This statement indicates a high degree of confidence in the Admiralty's own design and corroborates Professor Jack Edwards' opinion that the UK would have built its own reactor, independent of US influence, albeit would have taken some two years longer. 'It would not have been as good as *Skipjack* [...] but it would have been entirely of our own design and would not have made us so dependent on the whim of the US Congress on the passage of further information to us'.[139] Although Edwards' views are given with the hindsight of forty years, they do illustrate that not everyone associated with the project was keen to see their hard work

[135] TNA, AB 8/791, Minute, White, 18 September 1958.

[136] TNA, AB 16/2742, United States of America – United Kingdom collaboration on submarine reactor, 1958–59, Letter, Gaunt to Strath, 22 August 1958.

[137] TNA, DEFE 19/50, Note, RANP 21/1/58, 21 January 1958.

[138] TNA, ADM 1/27375, Nuclear submarine and POLARIS Project, 1957–58, Memorandum, Controller to Baker, 6 August 1958.

[139] Private correspondence, Jack Edwards to Vice Admiral Sir Robert Hill, 10 April 1998.

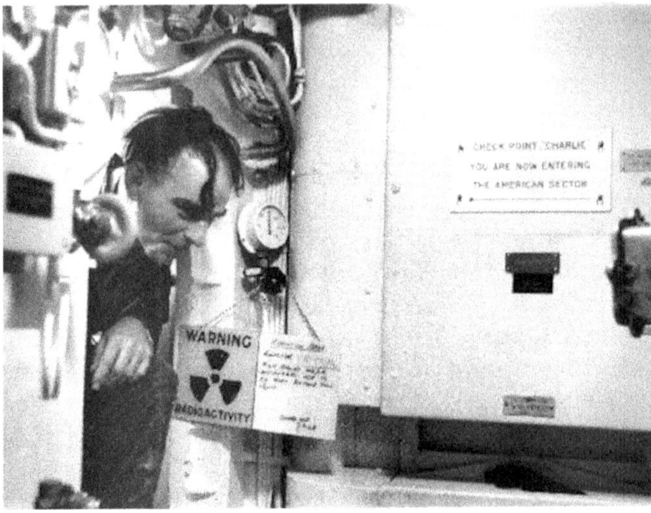

Fig. 5.2 L/Mech King entering the reactor tunnel in HMS *Dreadnought* (note the checkpoint Charlie sign)

apparently abandoned in favour of the American S5W plant and chimes with Crowe's observations concerning opposition to the purchase of S5W plant. It is apparent from the memorandum, however, that the designers were struggling with certain aspects of the core design and a safe and reliable control mechanism for the reactor; hafnium was a rare metal and in short supply in the UK. This was possibly one of the reasons why Mountbatten had pushed for transfer of American technology so it would advance HMS *Dreadnought*'s commissioning.

Rolls-Royce was appointed as the Admiralty's agent 11 August 1958 and two days later was authorised to open negotiations with Westinghouse to purchase the S5W plant and associated machinery.[140] So that the US nuclear propulsion programme was not interrupted, Rickover required Westinghouse to establish a separate organisation known as the Westinghouse Special Atomic Programme (WSAP). One of the most contentious points was Rickover's absolute refusal to allow Rolls-Royce to sub-contract the manufacture of the fuel elements; in effect this meant the UKAEA had the UK monopoly on their production. The UKAEA released

[140] DNP 2, NP184/2011, Letter, C.P. 8B/58380/58, 11 August 1958, p. 46.

the Admiralty from its obligation to have them manufacture the fuel elements should: '...the Admiralty want to make other arrangements'.[141] Lang wrote to Sir Norman Brook, the Cabinet Secretary, to forewarn him that because of Rickover's insistence on Rolls-Royce manufacturing the fuel elements, there would need to be a Ministerial decision on the future policy for fuel element manufacturing.[142] The Admiralty asked the UKAEA to give an assurance that any information passed to them from Westinghouse would only be used for defence purposes. The Atomic Energy Executive discussed whether Westinghouse information would prevent them from certain commercial activities and decided that if Rolls-Royce manufactured the submarine fuel elements the Authority would not require this information from Westinghouse.[143] After much negotiation, a draft supply contract and licensing agreement had been drawn up with a break-down of costs totalling more than \$29 M. This equated to a little more than £10 M which was higher than the original £8 M authorised by the Chancellor in February and resulted from the decision to proceed with the S5W rather than S3W reactor and to purchase compatible US propulsion machinery.[144] The draft agreement initially allowed fuel elements to be manufactured by others after three years, effectively ending Rolls-Royce's monopoly. Lang recommended that the Government endorse the Agreement and Rolls-Royce be allowed to manufacture fuel elements.[145] The Government accepted that the Admiralty should agree to the draft contract, the Admiralty signified their acceptance by forwarding a letter of intent to Rolls-Royce advising that a contract would be placed once the Westinghouse/Rolls-Royce contract and Licensing Agreement had been endorsed by the Government.[146]

When Rickover received the supply contract and Licensing Agreement, he objected to the Licensing Agreement stating no one, even he, had the authority to negotiate a Licensing Agreement and would not consider the supply contract until the Agreement was withdrawn.[147] The Licensing Agreement with Westinghouse would have been important for two

[141] TNA, AB 16/2742, Telegram, PEW 2337, 2 October 1958.

[142] TNA, CAB 21/4852, Letter, Lang to Brook, 22 October 1958.

[143] TNA, AB 16/1380, Extract from A.E.X.(58), 13 November 1958.

[144] TNA, CAB 21/4852, Memorandum, D.(58)72, 20 November 1958.

[145] TNA, CAB 21/4852, Memorandum, Lang to Brook, 20 November 1958.

[146] TNA, T 225/1192, Development of a nuclear submarine [...], 1958, Letter, 8B/62209/58, 5 December 1958.

[147] DNP 2, NP184/2011, US Objection to License Agreement, p. 51.

reasons; firstly, exchange of information would have continued for the ten years the supply contract was in force; thus, the Admiralty would get information on improvements in the core design or details of any problems encountered. Secondly, Westinghouse would have conducted certain research, development and testing for Rolls-Royce on repayment.[148] Selkirk wrote to Macmillan to advise him that the US had insisted on excising large parts of the Licensing Agreement, under which future reactor development information would have passed to the UK. Although this was a huge disappointment, Selkirk advised that enough remained of the Agreement to make it worthwhile. Selkirk noted that renegotiation was an option to allow future collaboration on reactor development; however, it was not an opportune time as Rickover was due to visit within a few days and Selkirk wanted to be able to tell Rickover that the Government agreed to the contract whilst he was in London.[149] In reply, Macmillan expressed disappointment that the possibility of future reactor collaboration was much reduced, no doubt it was not what he envisaged for the policy of interdependence but he agreed to the proposed changes on the understanding they were accepted by the US Authorities.[150]

One major advantage of withdrawing the Licensing Agreement was Rolls-Royce would be entitled to manufacture future naval reactors on information received under the supply contract without payment of royalties as previously envisaged, thus enabling substantial future savings in dollar expenditure. Rolls-Royce and Westinghouse reached agreement 10 February 1959, which was acceptable to the Admiralty and US Authorities. The following day Macmillan approved the contract, and the Presidential determination, required under the 1954 Atomic Energy Act, was made 5 March.[151] For their money the Admiralty would receive one S5W reactor with a Type 2 core including zircaloy fuel elements and hafnium control rods with the associated machinery sited in the reactor compartment. The engine room equipment included main turbines, turbo-generators and a comprehensive list of auxiliary machinery. The total cost, which included technical and manufacturing assistance, training, shipment, etc. amounted

[148] DNP 2, NP184/2011, US Objection to License Agreement, p. 51.
[149] TNA, PREM 11/2635, Letter, Selkirk to Macmillan, 28 January 1959.
[150] TNA, T 225/1193, Note, Macmillan to Selkirk, 9 February 1959.
[151] DNP 2, NP184/2011, Agreement Concluded, Contract Signed, p. 51.

to $29,243,369 (roughly £10,450,000) which was spread over six years, 1959–1964.[152]

Article II of the Agreement allowed for the exchange of information and paragraph five in particular related to the: 'research, development and design of military reactors to the extent and by such means as may be agreed'.[153] However, although the Admiralty saw this paragraph as having the potential for a much wider scope of transferring nuclear information, it proved in practice not to be the case. Information flowed through the Westinghouse/Rolls-Royce contract under the authority of the USAEC and was essentially confined to design and plant instruction-type information. Despite attempts to open further discussions with the US, the Admiralty was rebuffed by Rickover who held a rigid interpretation of the Mutual Defence Agreement focussed on Article III, the Bilateral Treaty: '...by which Britain was restricted to purchase one set of Reactor and Propulsion Machinery only for DREADNOUGHT and "know-how" to permit design and manufacture in the future'.[154] Writing some years later, Captain E. P. C. Kelly observed that paragraph five had boiled down in practice to the Westinghouse/Rolls-Royce contract (a new contract between Rolls-Royce and Associates and Electric Boat Division of General Dynamics covered the work linked to *Dreadnought*'s first refuelling). 'It has become evident over the years that this is the limit and with the completion of DREADNOUGHT'S refuelling, the flow of information ends'.[155] The intentions of "by such means as may be agreed" written in paragraph five of Article II would never meet the Admiralty's expectations as it would never be countenanced by Rickover.

Final Adjustments

The combined contractors' team that had been set up at Derby in the autumn of 1957 consisting of engineers from Rolls-Royce, Vickers-Armstrongs and Foster Wheeler formed the nucleus of a new company, Rolls-Royce and Associates Ltd. Formed 6 February 1959, Rolls-Royce had the controlling stake in the new company with fifty-two percent of the

[152] TNA, T 225/1191, Development of nuclear submarine, including purchase of machinery from USA: fissile material programme on nuclear propulsion for staff, 1958, Letter, M.I.618/2/58, 20 November 1958.

[153] Mutual Defence Agreement, 1958, p. 3.

[154] DEFE 72/45, Letter to Baker, 19 March 1961.

[155] DEFE 72/45, Memorandum, ADME/NP/113/67, 25 April 1967.

shares; Vickers-Armstrongs and Foster Wheeler held the remaining shares equally.[156] Baker called a meeting to clarify the new company's functions so contracts for HMS *Dreadnought* and her machinery could be finalised. It is clear from the Minutes that there was disagreement between the parent companies over Rolls-Royce and Associates' functions. Whilst it was agreed that Rolls-Royce and Associates would not manufacture any equipment, the Vickers-Armstrongs representatives were adamant that Rolls-Royce and Associates should be a design company only whereas the Rolls-Royce representatives argued that Rolls-Royce and Associates would also have a procurement function. Previously, Vickers Nuclear Engineering Limited had managed cooperation between Barrow and Derby and the question was asked who, now, would coordinate the work. Baker wrote: 'The meeting then degenerated into a coalition of intransigence'.[157] A further meeting was held at which the three parent companies agreed to a procurement function for Rolls-Royce and Associates, it was envisaged that the Admiralty would place a contract with the shipbuilder, Vickers-Armstrongs, who would be instructed to place a sub-contract with Rolls-Royce and Associates for the reactor compartment. Rolls-Royce and Associates in turn would place sub-contracts for the detailed design and procurement of the main components of the reactor compartment with the parent companies as far as the Admiralty would allow.[158] The building of a submarine is a complex undertaking and a nuclear-powered submarine more so; therefore a further memorandum sought to clarify the relationship between the companies. Vickers-Armstrongs (Shipbuilders) was contracted to build the submarine; Rolls-Royce held the contract to obtain the reactor and propulsion machinery from the US and proposed to use Rolls-Royce and Associates as their managers. Vickers-Armstrongs (Engineering) was sub-contracted to install the machinery whilst Rolls-Royce and Associates and Westinghouse retained technical engineering control of the installation.[159] In place of Westinghouse, at Dounreay, Foster Wheeler was responsible for all the engineering components of the reactor compartment excluding the core and control rods. However, disagreement over Vickers-Armstrongs responsibilities and Rolls-Royce and

[156] Lambert, *Rolls-Royce: the Nuclear Power Connection*, p. 33.

[157] DNP 5, N/CP24, Rolls-Royce & Associates Ltd., Constitution, Letter, RB/49/59, 10 February 1959.

[158] DNP 5, N/CP24, Letter, Dunphie to Baker, 12 March 1959.

[159] DNP 5, N/CP24, Memorandum, C.P.52769/59, 3 June 1959.

Associates' responsibilities continued for some time; at Dounreay the Admiralty thought Rolls-Royce and Associates should be selected as contractor with the Chief Engineer a member of that company. It was expected Rolls-Royce and Associates would sub-contract to Vickers-Armstrongs (Engineering) the operation of the plant and site machinery.[160] Whilst there may have been some ill-feeling at Rolls-Royce and Associates gaining the Westinghouse contract to the detriment of Vickers-Armstrongs, from the correspondence researched, these disagreements must be viewed as a means of delineation and clarifying each company's responsibilities.

The industrial element of the programme had been addressed as had the Royal Navy organisation with the formation of the DPT. One other area to attend to was oversight of the supply contract in the US. Captain D. A. Cotman and a team of fifteen naval engineers, constructors and draughtsmen were appointed to HMS *Saker*, Washington D.C., to oversee the transfer of technical information and inspect manufactured equipment coming off the production line. As Dreadnought Liaison Engineer, Cotman and his team would work from Westinghouse offices at Pittsburgh, Pennsylvania and Electric Boat offices at Groton, Connecticut.[161] The majority of the DLE team, with Rolls-Royce and Associates staff, were based at Pittsburgh; Westinghouse was providing the reactor plant and the propulsion machinery. A smaller DLE team, with staff from Vickers-Armstrongs was based at Groton. Roger Berry had responsibility for overseeing all mechanical equipment procured for HMS *Dreadnought* by Electric Boat that did not form part of the main propulsion or reactor plant package. Other DLE members were Dick Brown (Electrical) and Mr. K. Foulger (Construction).[162] Apart from overseeing the procurement side of the contract, the DLE team at Groton was also responsible for transferring the vast amount of supporting documentation to the DPT. Colin Farley-Sutton could be described as Berry's opposite number for it was he, in DPT 2 section, that received the technical information from Berry pertaining to practically all the mechanical items not forming part of the propulsion system and reactor plant. Farley-Sutton's job was to interpret US naval and industrial details and practices into UK systems and practices, so they could be used for ship's drawings, handbooks and other

[160] TNA, AB 6/2308, Nuclear submarine (Dreadnought) project organisation, 1959–60, Letter, C.P.52769/59, 29 March 1960.

[161] DNP 2, NP184/2011, Letter, C.E.5063/58, 7 November 1958, p. 59.

[162] Private papers, Commander Roger Berry RN Ret'd.

documentation such as maintenance schedules.[163] The Royal Navy was just one of the organisations involved in the Dreadnought programme receiving technical information transferred from the US; *Dreadnought*'s crew also required reliable and accurate information for training, operation of the equipment, maintenance and fault finding.

In March 1959, the first of sixteen officers and thirteen senior rates were appointed to HMS *Dreadnought*. Lieutenant Commander B. F. P. Samborne was appointed Commanding Officer, Lieutenant Commander Peter (Spam) Hammersley was appointed Engineering Officer and Lieutenant Commander R. R. Squires, First Lieutenant. The keel for HMS *Dreadnought* was laid 12 June 1959; the fact that Prince Philip performed the keel laying ceremony (an unusual event for a member of the Royal Family to perform) reflected the importance the Government now attached to nuclear propulsion and, after much initial secrecy, was a means of raising awareness of submarine nuclear propulsion to the general public.[164] The nuclear propulsion project by mid-1959 was well organised, the DPT had settled into their work and there was one final organisational matter to attend to. With Baker now in control of the DPT, the post of Rear Admiral Nuclear Propulsion was discussed and was dissolved upon Wilson's retirement in December 1959. It was decided that future responsibility for nuclear propulsion lay with Director General Ships (DG Ships), Rear Admiral R. S. Hawkins.[165] By February 1961, it was reported that Electric Boat and Westinghouse had more than fulfilled their contractual obligations and delivered the majority of the machinery which had already been landed on the submarine, the main part to be delivered was the core (Fig. 5.3).[166]

In more ways than one HMS *Dreadnought* was a first in many respects. Her bow-mounted sonar array, Type 2001, was more advanced than any sonar yet developed and was a priority espionage objective of the Soviet Union.[167] A Water Ram Discharge system for firing torpedoes was also highly advanced; both these systems were to be fitted without the usual prototypes and trials which could have delayed completion by a further

[163] Private papers, Captain Colin Farley-Sutton RN Ret'd.

[164] 'Atom submarine to be built quickly', *The Times*, 13 June 1959, p. 8, Issue 54486.

[165] TNA, AB 16/1380, Letter, C.E.1820/59/A, 15 January 1960.

[166] TNA, AB 8/966/1, Part 1, Marine reactor development: Dreadnought programme, 1959–63, Letter, Baker to Kinsey, 3 February 1961.

[167] 'Police in secrets case describe finds', *The Times*, 9 February 1961, p. 7, Issue 55001. For further details see articles and books on the "Portland Spy Ring".

Fig. 5.3 HMS/m *Dreadnought* first Commissioning

eighteen months.[168] HMS *Dreadnought* was far in advance of any submarine, possibly even the *Skipjack* class; Professor Louis Rydill, designer of *Dreadnought*'s hull, has written of the five milestones in submarine design. First was the Holland, tested for the US Navy in 1899; the second milestone was the German U35 class with the emphasis on surface performance to allow surface transit using diesel engines, as briefly discussed in Chap. 1. The third milestone was Nazi Germany's type XXI with the emphasis now on streamlining and increased battery size for improved submerged performance. The fourth milestone was the USS *Albacore* hull form, a radical redesign of the pressure hull with a length to beam ratio of 7:1 (previous designs of submarines tended to have a length to beam ratio of 11:1). The final milestone was the USS *Skipjack*, the result of combining the *Albacore* hull with nuclear power.[169] As noted, only six nations

[168] DNP 2, NP184/2011, Vice Controller's memorandum, B.1275, 13 July 1959.
[169] Roy Burcher and Louis Rydill, eds, *Concepts in Submarine Design* (Cambridge, Cambridge University Press, 1994), pp. 12–20.

have reached the fifth milestone in submarine design. Finally, as if to emphasise the importance of HMS *Dreadnought* to the nation, she was launched by HM the Queen at Barrow and to underline her significance to the Royal Navy the date chosen was 21 October, *Trafalgar Day*.

By the end of the following year, despite numerous strikes threatening to delay her commissioning, HMS *Dreadnought* was complete and on 12 November 1962, her reactor went critical for the first time.[170] The following month HMS *Dreadnought* was ready for sea trials; however, with a sensationalist headline, *The Times* reported that HMS *Dreadnought* was delayed sailing. Although the loss of "primary" communication with the bridge can be a problem, the possible real reason for delaying sea trials was at the bottom of the article which reported a contributory factor to the postponement was heavy weather. 'Heavy seas were breaking over the tugs and a 45 m.p.h. wind was blowing from the south-west'.[171] HMS *Dreadnought* sailed on her maiden voyage in Morecombe Bay the same day as *The Times* report, 12 December 1962; in April she successfully completed her sea trials and was handed over to the Royal Navy at sea at a depth of 150 ft.[172] HMS *Dreadnought* was commissioned into the Royal Navy at Barrow 17 April 1963; finally, the Royal Navy had its own nuclear-powered submarine and a fleet was following. HMS *Valiant* was in build and HMS *Warspite* was on order.

The "special relationship" between the US and UK is rarely understood, badly reported, and often ridiculed in parts of the press and media. However, despite poor reporting and a lack of understanding, the US/UK "special relationship" is more than the portrayal of the US President as the leader of the western world with a lickspittle British Prime Minister pandering to his whim. The Mutual Defence Agreement was the culmination of Macmillan's foreign policy objective of achieving interdependence with the Americans whereby nuclear technologies and information would be exchanged, pooling research and development as equal partners. The Agreement came about against the geo-political backdrop of the Soviet Union's invasion of Hungary, demonstrating Soviet determination to militarily intervene in the home politics of countries under its sphere of

170 Desmond Wettern, *The Decline of British Seapower* (London, Jane's Publishing Company Limited, 1982), p. 211.

171 'Blown fuse holds up Dreadnought', *The Times*, 12 December 1962, p. 6, Issue 55572.

172 CND Archive, CND/1/68, Polaris, 1962–68, Pamphlet, Vickers Limited, Annual Review 1963.

influence, and the demonstration of its technological lead in ballistic rockets with the launch of *Sputnik*. Macmillan had been keen to restore US/UK relations after the debacle of the Suez crisis and in Eisenhower he found a willing partner who shared his objective. John Baylis wrote of the "special relationship" that successful negotiation of the MDA was not only down to the roles of Eisenhower and Macmillan, but also what he called a "transatlantic advocacy coalition" consisting of nuclear scientists, defence officials and intelligence officers. Baylis argued that these men: '…also played a part at the operational level in achieving and subsequently shaping the kind of relationship which developed'.[173] Michael Goodman has also written on the subject of Anglo-American relations and has gone further noting that atomic intelligence exchanges continued after World War II and these developed into a relationship, based on a ***modus vivendi*** which was signed 7 January 1948. This agreement established nine areas of collaboration from which, Goodman argues, all further developments in exchanging nuclear information can be traced. 'The achievement of Macmillan's "great prize" therefore, was based on the foundations established through atomic intelligence'.[174]

This chapter has demonstrated that personal relationships mattered greatly. Macmillan and Eisenhower had worked together during World War II, but the Mutual Defence Agreement was/is more than a political relationship as Baylis argued. Mountbatten enjoyed a very close relationship with his opposite number in the US Navy, Admiral Burke USN, which helped immensely with the smooth introduction of Polaris to the Royal Navy. However, the one relationship to have had the most dramatic impact on the Admiralty's nuclear propulsion programme was that of Admiral Rickover and Mountbatten, of which it has been written that: 'The introvert iconoclast from the Ukraine fell under the spell and aura of Queen Victoria's grandson'.[175] Whatever the language used to illustrate their relationship, it should not be underestimated, especially in the context of the UK's nuclear submarine programme. There is much more entwining the US and UK than a common language; in his introduction to *A History of the English-Speaking Peoples Vol 1*, Churchill described them as: '…those independent nations who derive their beginnings, their speech, and many of their institutions from England, and who now

[173] Baylis, 'The 1958 Anglo-American Mutual Defence Agreement:', p. 425.
[174] Goodman, 'With a little help from my friends:', p. 175.
[175] Ziegler, *Mountbatten*, p. 558.

preserve, nourish and develop them in their own ways'.[176] Given so much commonality between the UK and US and the deep military and nuclear collaboration of World War II, it was, perhaps, inevitable that a "special relationship" of some sort would develop between the two countries. With the onset of the Cold War—a mainly submarine war—NATO would rely heavily on US Navy and Royal Navy nuclear-powered submarines to keep the peace.

[176] Winston S. Churchill, *A History of the English-Speaking Peoples Vol. 1: The Birth of Britain* (Electronic edition New York, Rossetta Books, 2013).

CHAPTER 6

Nuclear Training and Dounreay

Prior to getting HMS *Dreadnought* to sea, there was much work to be done in relation to the educational requirements and training of personnel. This chapter will focus attention on the academic and practical training Royal Navy officers and ratings received to equip them with the necessary information and skills to operate a pressurised water reactor power plant and associated machinery. The Royal Navy was at the forefront of employing this new technology and had to design its own training requirements, these were closely aligned to US Naval experience and practices. It would not only be the officers and ratings that required training but also Admiralty civilian staff. The firms Vickers-Armstrongs, Rolls-Royce, Foster Wheeler and Rolls-Royce and Associates would all require some of their staff training on different aspects of manufacturing techniques. Finally, dockyard workers who would be employed in de-fuelling and refuelling nuclear submarines would also receive special training in the procedures for handling the core.

One major problem caused by the purchase of the S5W plant was that HMS *Dreadnought* would be completed before the Dounreay Submarine Prototype (DS/MP). One of the main functions of the Dounreay prototype was for training Royal Navy personnel on operating the reactor using Standard Operating Procedures (SOPs) and being able to deal with any crisis with Emergency Operating Procedures (EOPs). This important

G. M. Jones, *The Development of Nuclear Propulsion in the Royal Navy, 1946–1975*, Security, Conflict and Cooperation in the Contemporary World,
https://doi.org/10.1007/978-3-031-05129-6_6

training facility would not now become available until sometime after HMS *Dreadnought*'s commissioning, and with no plant to allow practical training of personnel in the UK, the Royal Navy would initially be dependent on the US Navy for training the first cohort of its nuclear engineering submariners. This chapter will study the importance of Dounreay's role as the prototype plant for evaluating maintenance procedures and techniques, and as a test platform for auxiliary and secondary machinery and for testing new cores. Questions were raised by the Treasury over the expense of Dounreay and whether the Royal Navy required a prototype plant now that it was purchasing a S5W reactor plant from the US. However, Dounreay would quickly prove its value and importance during its commissioning.

Training Facilities

Whilst the Admiralty nuclear submarine project had been in existence since early 1946 and had been substantially expanded since 1954, it was not until 1956 when attention finally turned towards training the personnel that would operate and maintain the nuclear reactors being fitted into submarines. The Controller wrote to Mountbatten advising consideration be given to the offer of training facilities in the US for engineer officers and the formulation of a training policy to take advantage of various courses available at Harwell and at Oxford and Manchester universities, with the aim of creating a pool of trained officers and ratings.[1] At that time, Royal Navy officers, Admiralty staff and members of Vickers-Armstrongs and Rolls-Royce joining the Naval Section at Harwell were attending the reactor school there as part of taking up their duties; the training was mostly connected to gas-cooled, graphite-moderated reactors. In the spring of 1957, a memorandum was produced to discuss the scope of the exchange agreement within which it questioned the possibility of Royal Navy personnel training in US Navy establishments. Initial discussions were held with interested departments and some ideas formed of the likely format that training should take. However, following the Technical Mission's visit to the US in June 1957, it was decided from the information obtained of the US Navy's training methods that the Royal Navy's ideas would have to be radically developed. Wilson produced a paper in which his proposals were discussed at a meeting in December 1957; there was a consensus that further study of the US system was

[1] TNA, ADM 205/112, Letter, Controller to Mountbatten, 6 September 1956.

required before a firm policy could be submitted.[2] A visit by training experts to the US was in the process of being planned when Rickover's offer to the UK to purchase a US reactor was made; a view was taken that rather than prejudice the early negotiations, no further action would be taken for the time being. At a meeting attended by the Second Sea Lord, the senior officer in charge of personnel matters, 25 April 1958, a decision was made to proceed with formatting a training policy without further recourse to US training methods.[3] The Royal Navy was slowly losing time to train its engineers and it was better to have some form of training policy in place rather than none.

With the formal offer to purchase a US Naval propulsion reactor made in January 1958, it would soon become obvious that if the Royal Navy took up the offer, their training regime would not be sufficient to get the officers and ratings qualified in time for HMS *Dreadnought*'s launch and sea trials. It was recognised in the current circumstances that the US was not able to assist with training and the Admiralty would examine what training facilities could be organised in the UK; however, they may have to raise the matter with the US Navy later if it proved necessary.[4] When the Mutual Defence Agreement was signed there was no reference to US assistance in training HMS *Dreadnought*'s crew, despite efforts to amend paragraph A of Article III. As far as the US Government (Rickover) was concerned, it was quite simply a transfer agreement. Once again, Mountbatten would have to ask favours of his counterpart, Admiral Burke USN. During 16–20 October 1958, Mountbatten held discussions with Burke, which at some point also included Rickover and the Bureau of Personnel Chief, Vice Admiral H. P. Smith USN. At this meeting it was agreed that Rickover and Smith should put some proposals together as to how they could assist the Royal Navy in training an adequate number of engineering personnel for HMS *Dreadnought*. It has to be understood that although all navies will require various professional branches such as engineering, executive and supply to man their ships, the educational standards required and the training given will differ with each navy as to the equipment fitted and depth of knowledge required to operate/maintain it. There is also a financial question of affordability when deciding on training requirements. There were certainly discussions on the subject

[2] DNP 2, NP184/2011, Training of Personnel, p. 68.
[3] DNP 2, NP184/2011, Training of Personnel, p. 68.
[4] TNA, DEFE 13/523, Letter, Caccia to Dulles, 3 March 1958.

before selecting which branches of the Royal Navy would be best suited to receive US training. As Mountbatten noted: 'It would be a great disaster if we made a mess of your lovely equipment through failure to understand how to work it'.[5] Yet again, this tripartite relationship reaped rewards for the Royal Navy. Rickover wanted to assist in making HMS *Dreadnought* a success and saw little difficulty, if authorised by Burke, in training Royal Navy personnel in US Navy submarines.[6]

On his return, Mountbatten lost no time and wrote to Vice Admiral Smith USN to outline the problems he now envisaged by completing HMS *Dreadnought* ahead of Dounreay and the possible embarrassment that could ensue if the Royal Navy commissioned HMS *Dreadnought* but did not have the personnel qualified to take her to sea. Mountbatten asked Smith for the US Navy's complete training programme for officers and ratings, including syllabuses and specimen examination papers.[7] A short while later, Mountbatten received a letter advising him that Rickover had arranged with Vice Admiral Smith to take ten Royal Navy personnel for practical training onboard US Navy submarines.[8] However, the number of personnel that required training exceeded by some margin the training billets initially offered; a minimum of sixteen officers and thirteen senior rates appointed to HMS *Dreadnought* would eventually travel to the US to receive practical training in US Navy submarines.[9] In anticipation of Royal Navy personnel receiving practical training in the US, a meeting was held to discuss the educational standards and requirements of the prospective students.[10] The Director Naval Education Services, Instructor Rear Admiral J. Fleming noted the high academic standard of US Naval ratings attending the Nuclear School at New London, Connecticut, and it was agreed that only Marine Engineering Artificers were of an educational standard that would benefit from training in the US. It was envisaged that seaman officers would require a twelve-month course to bring them up to the academic and technological standards of the equivalent US naval officer, it was also advised that engineering officers should be of either

[5] Broadlands Archive, MB1/I447, Letter, Mountbatten to Burke, 26 October 1958.

[6] Broadlands Archive, MB1/I478, Report of First Sea Lord's Visit to Canada and the U.S.A., p. 30.

[7] TNA, ADM 205/178, Letter, Mountbatten to Smith, 3 November 1958.

[8] TNA, ADM 1/27198, Letter, Thistleton-Smith to Mountbatten, 14 November 1958.

[9] DNP 2, NP184/2011, Principal Officers Appointed, p. 84.

[10] TNA, ADM 1/27088, Nuclear propulsion: Naval personnel training, 1958–59, Minute, RANP.14/11/58, 20 November 1958.

"dagger" or honours degree standing.[11] The meeting recognised that due to the timescale, HMS *Dreadnought*'s key personnel would have to start their training in January 1959. Wilson also noted that educational standards would have to be raised; given the trend towards nuclear power, it was essential that the scientific standard of naval cadets be raised without delay and that all junior officers receive a thorough training in scientific subjects.[12] Apart from the millions being spent on HMS *Dreadnought*, the purchase of the S5W plant and the building and infrastructure of the Dounreay prototype, the Admiralty would also have to allocate some capital expenditure on the shore, training establishments to support the additional training requirements.

To train the personnel required to man the envisaged fleet of twelve submarines, it was not just a case of preparing courses and examinations, there would also be a requirement to build laboratories, classrooms and, later, simulators as well as accommodation messes and other ancillary offices. In the early days of the nuclear programme, naval officers and scientists who were seconded to the Naval Section at Harwell attended civilian training courses at Imperial College, Queen Mary College and Harwell. It became apparent that these courses were not going to meet the Navy's requirements; the courses concentrated on gas-cooled, graphite-moderated reactors and focussed on design principles, they could not offer operational experience. The courses were also limited by the number of places available and their timing; in particular the colleges could not meet the security requirements of the submarine nuclear plant, as required by the MDA to protect naval nuclear propulsion information; it was evident that the Royal Navy would have to train its own people.[13] One further aspect, unique to the submarine reactor plant that required addressing, is the fact that the submarine would operate away from support services that would normally be available to land-based reactor operators. It was evident that the men responsible for operating a nuclear power plant under those circumstances would: '…have to possess such a thorough understanding of the theory underlying the plant's behaviour, that they are capable of both

[11] The "Dagger" was a typographical symbol used in the *Navy List* to denote officers who had completed further courses in their specialisation.

[12] TNA, ADM 1/27088, Register No. NCW/429/2/58/B, Minute Sheet No. 3, para. 4, 9 February 1959.

[13] J. Edwards, 'Nuclear Courses at the Royal Naval College', *Journal of Naval Engineering*, Vol. 19, Book 1, (Jun., 1970), 15–24 (p. 16).

reasoned and instinctive action in all circumstances'.[14] The correct initial actions in any emergency are the difference between a disaster occurring and a controlled evolution to revert a system to normal operation—especially in a submarine.

Off to School

The first bespoke training for Royal Naval officers was an introductory course of two weeks duration, intended for officers of all specialisations taking up appointments in the nuclear field. Announced in Admiralty Fleet Order, AFO 2022/58, 22 August 1958, the first course started 20 October at HMS *Collingwood*, the Royal Naval Electrical School at Fareham.[15] Soon afterwards, the Department of Nuclear Science and Technology was formed at the Royal Naval College, Greenwich, in January 1959. Professor Jack Edwards, the senior RNSS scientist at Harwell since January 1955, was appointed to the Chair of the new Department and would initially divide his time between Greenwich and Harwell to clear up outstanding work from the run-down of the Naval Section.[16] The Department at Greenwich was established to educate naval officers in nuclear science and technology with the remit to keep in touch and be up to date in nuclear technology developments. Although it would be four years before HMS *Dreadnought* would commission, Edwards noted that the timing of the first courses was governed by the requirement to produce academically qualified officers in nuclear technology, so they were ready to accept the practical training opportunities in US Navy nuclear submarines prior to taking up their own appointments to stand-by submarines under construction.[17] However, with the purchase of the S5W plant, HMS *Dreadnought* was due to commission two years earlier than anticipated and this would have an immediate impact on the Admiralty's training programme. Based on a report by a fact-finding mission to the US to look at training in January 1959, the Admiralty decided in February that HMS *Dreadnought*'s crew must be available to go to the US in two groups. The first would commence training in April, with the second group

[14] Edwards, 'Nuclear Courses at the Royal Naval College', p. 16.

[15] DNP 2, NP184/2011, Training of Personnel, p. 66.

[16] TNA, AB 6/1599, Liaison with Admiralty Research Laboratory [...], 1955–59, Summary, Edwards, 18 December 1958.

[17] Edwards, 'Nuclear Courses at the Royal Naval College', p. 15.

starting in October. In doing so the Admiralty recognised the undesirable fact that the officers and ratings selected would not receive a satisfactory level of nuclear training before their departure.[18] Apart from these two groups the Admiralty was confident that the remaining personnel would receive a reasonable level of academic and technical instruction prior to their practical training in the US. It is worth commenting here that Rickover interviewed and selected all officers entering the US Navy nuclear submarine programme (some 14,000 personnel over a thirty-year period) and, in order to maintain the highest standards, was keen to extend his influence to cover the Royal Navy's selection process for aspiring nuclear submariners. However, his offer was politely declined by Mountbatten as constitutionally impossible. 'Can you imagine the scene in Congress if a British Admiral were selecting U. S. N. personnel…'?[19]

The training these early nuclear engineering officers received is typified by the experiences of Lieutenant Commander Peter Hammersley (later Rear Admiral). Hammersley was appointed as the first Engineering Officer of HMS *Dreadnought* and has written of his training as one of the Royal Navy's first nuclear engineers. As noted previously, Hammersley was appointed to HMS *Dreadnought* in March 1959; however, he had been advised of this appointment in 1957. Hammersley left his appointment in the conventionally powered submarine, HMS *Tiptoe*, and attended Imperial College, London, in August 1958 for a one-year post graduate course in nuclear engineering where he and the other students learnt a lot about gas-cooled, graphite-moderated "nuclear power stations", Hammersley recalls he received an excellent grounding in nuclear technology upon which he was able to build. On completion of his studies at Imperial, Hammersley (with CERA T. Faulkner and CPO Mechanician I. Maryon) was appointed in October 1959 to the USS *Skipjack* based at Groton, Connecticut. In the US, Hammersley and the two Chief Petty Officers would receive practical training and experience on a S5W submarine reactor. The designated Commanding Officer of HMS *Dreadnought*, Lieutenant Commander B. F. P. Samborne, had been appointed to the USS *Skipjack* in May 1959 and had already completed his training. Comments by Samborne which may have had an effect on HMS *Dreadnought* were forwarded to FOS/M and Director General Ships in a

[18] DNP 2, NP184/2011, Training in the U.S.A., p. 72.
[19] TNA, ADM 1/27198, Letter, Mountbatten to Rickover, 13 January 1959.

report, SM.169/14/11/M, 16 December 1959.[20] Royal Navy officers, such as HMS *Dreadnought*'s newly appointed First Lieutenant, Lieutenant Commander R. Squires, and ratings that were appointed to the USS *Skate* and other submarines of that class, would receive their training on the S3W reactor or S4W reactor plant, whichever was fitted.[21] Rickover had decreed that Royal Naval officers should not "keep watch" on any power plant except under supervision of a qualified US Navy officer. The Commanding Officer of the USS *Skipjack*, Commander Bill Behrens, said that if he was satisfied that Hammersley was competent, he could take charge of watches on *Skipjack*.[22] After five months understudy, Hammersley satisfied Behrens and qualified; not only did he keep watches, Hammersley also had new US Navy officers understudying him, in direct defiance of Rickover's directive.[23]

With Hammersley at Imperial was Lieutenant Commander John Grove who would be appointed as the first Electrical Officer of HMS *Dreadnought*. On completing his studies at Imperial, Grove was appointed in October 1959 to the USS *Skate*, also based at Groton, to continue his training. In May 1960, Hammersley returned to the UK and began work at Barrow with HMS *Dreadnought* building. During their time with the US Navy, Hammersley and Grove realised that although it was normal in those days to be appointed Engineering Officer and Electrical Officer, respectively, it would be unworkable on a nuclear-powered submarine. The Engineering Officer was responsible for the reactor and the Electrical Officer was responsible for the reactor control systems. Both officers decided to combine the engineering and electrical departments on HMS *Dreadnought* with a departmental head and deputy. As Grove had six months seniority, he became the Senior Technical Officer and Hammersley became his deputy, the Marine Engineering Officer.[24] This organisation of the "Back-Aft" engineering department has continued to this day in Royal Navy

[20] DNP 2, NP184/2011, Principal Officers Appointed, p. 84.

[21] See appendix for list of Royal Navy Officers and Ratings trained in USN submarines.

[22] W. W. Behrens later became a Vice Admiral and Oceanographer of the US Navy. Upon retiring he assisted in setting up the National Oceanic and Atmospheric Administration, (NOAA).

[23] Private papers, Rear Admiral Peter Hammersley.

[24] Private papers, Rear Admiral Peter Hammersley, Speech to IMarEST Lunch Club, 17 September 2004.

submarines although the titles have slightly changed.[25] During his time at Imperial, Hammersley met Mountbatten twice and Rickover once in London; he met Rickover twice more during his time on the USS *Skipjack*. Hammersley considered that Rickover had done a great deal to help the British and was able to call on him in Washington to thank him before returning home. Hammersley formed the opinion that Rickover was a rude and difficult man but was convinced: '...that the Free World owes a lot to him for the preservation of peace throughout the time of the Cold War and I am glad that I went to see him'.[26] This is a view held by many who worked for and knew Rickover, including President Jimmy Carter, a former US Navy submarine officer who had endured Rickover's formidable interview process.

Initially two courses were run at Greenwich: the Nuclear Reactor Course of two terms duration and the longer Nuclear Advanced Course of three terms duration. Both courses are preceded by a term of the Nuclear Preparation Course which served as an academic refresher for mathematics, chemistry, thermodynamics and other subjects prior to advanced study of nuclear engineering.[27] It must be noted that some students selected for nuclear training would have already completed basic training at Dartmouth followed by an engineering degree or diploma at the Royal Naval Engineering College, Manadon, and joined Greenwich direct from successful completion of a submarine course at HMS *Dolphin*. These officers would have been in continuous education since starting school. Other officers, however, would have served in conventionally powered submarines or surface warships and, not having been in a formal educational environment for some time, may possibly have benefitted from the academic preparation. Unlike many university courses, Edwards recorded that the students at Greenwich had a diversity of professional experience and their ages ranged from twenty-three to thirty-five.[28] Training at Greenwich was also provided for personnel from other specialist Departments of the Admiralty such as the RNSS and the RCNC, as well as staff from private industry.

[25] The propulsion department is led by the Marine Engineering Officer (MEO) and his Deputy (DMEO).

[26] Private papers, Rear Admiral Peter Hammersley, Speech to IMarEST Lunch Club, 17 September 2004.

[27] DNP 2, NP184/2011, Introductory nuclear course, p. 69.

[28] Edwards, 'Nuclear Courses at the Royal Naval College', p. 17.

The Nuclear Advanced Course was designed for officers that would participate and ultimately take responsibility for a nuclear project (such as Admiralty scientists who would not serve onboard an operational submarine), in the case of Royal Naval officers this would extend to the safe operation of a submarine's nuclear plant. In due course, a Master of Science degree was awarded on completion of this course. The Nuclear Reactor Course was designed for officers who would have direct responsibility for the operation of the Royal Navy's nuclear power plants. This course's syllabus differed in that it concentrated on current nuclear power plants and their operational considerations.[29] A further three months of practical training would follow at the Dounreay prototype when it commissioned. The Department at Greenwich also provided a number of shorter classes; the nine-week Nuclear General Course was intended for seaman officers appointed to nuclear submarines and the first six weeks were deemed suitable for dockyard engineers with responsibility for maintenance and refuelling nuclear submarines. A three-month Nuclear Radiation Protection Course offered instruction in radiation monitoring equipment, health physics organisations and administration in health and safety matters. A Nuclear Introduction Course gave a simplified overview of reactor theory and plant principles which was suitable for officers in a support role.

Until the mid-1950s, ratings' mechanical training had been scattered around the Portsmouth area and farther afield at Chatham and Devonport, these included professional courses for the different mechanical trade branches of Artificer, Mechanician and Mechanic. Typically, a Mechanic would join the Royal Navy with little formal educational qualifications but would show an aptitude for engineering; an Artificer would join the Royal Navy, as an engineering apprentice, with educational qualifications in science and mathematical subjects; his career path would be one of rapid promotion to Petty Officer once his trade training was complete. The Mechanician was a Mechanic who was selected after his Leading Hand's qualifying course where his course grades merited further training to the same level as the Artificer and consequent better promotion prospects.

To consolidate all mechanical and professional courses in one area, HMS *Sultan* at Gosport opened as the Royal Navy's Mechanical Training Establishment 1 June 1956. With the nuclear submarine programme gathering pace, nuclear propulsion training was transferred in 1960 from

[29] Edwards, 'Nuclear Courses at the Royal Naval College', p. 18.

HMS *Collingwood* to what was now known as the Marine Propulsion Machinery School, HMS *Sultan*. Investment started to flow and in November 1961, the *Rutherford* building was opened to further facilitate nuclear training.[30] *Rutherford* building contained offices, classrooms and five laboratories for teaching metallurgy, water treatment, reactor instrumentation and control, physics and mechanics, and nuclear physics. Initially, two courses were established, the Nuclear Propulsion Long Course and Nuclear Propulsion Short Course; the Long Course was of twenty-three weeks duration and aimed at the technician level (Artificers and Mechanicians). This course was mainly comprised of theoretical training which would give the trainee sufficient knowledge and awareness of the reactor plant to enable them to progress to practical training at Dounreay. On completion of the Long Course, the technician would receive a further ten weeks of specialisation training. Electricians would attend HMS *Collingwood* for instruction in Health Physics instrumentation; some mechanical ratings would learn about water chemistry and some would attend a special welding course. The Short Course, as the name implies, was of ten weeks duration and was aimed at Mechanics and Sick Berth Attendants; the emphasis of this training was less theoretical and more plant based. On completion of this course the trainee would receive another week of specialisation training. Vickers-Armstrongs and Rolls-Royce and Associates took advantage of the courses being run at HMS *Sultan* sending staff for training; eventually Cammell Laird and the dockyards would also send staff there for instruction.[31] Reactor Panel Operators and Nuclear Chiefs of the Watch had to pass the Long Course before progressing to Dounreay for Nuclear Submarine Operator Training. This training consisted of eight weeks classroom and five weeks watch keeping, other ratings watch keeping aft would have to pass the Short Course prior to practical training at Dounreay. Their training at Dounreay comprised five weeks classroom and three weeks watch keeping (Fig. 6.1).[32]

The money that the Admiralty was investing in the men's training would require them to commit to a number of years' service. It was envisaged that Artificers, Mechanicians and Sick Berth Attendants serving in

[30] C. S. McHugh, 'H. M. S. 'Sultan' 25 Years of Marine Engineering Training', *Journal of Naval Engineering*, Vol. 26, Book 2, (Jun., 1981), 167–85 (p. 171).

[31] G. F. A. Trewby, 'H. M. S. Sultan: The Marine Propulsion Machinery School', *Journal of Naval Engineering*, Vol. 14, Book 3, (Dec., 1963), 315–33 (p. 328).

[32] M. A. Wilson, 'Nuclear Submarine Reactor Plant – Practical Training', *Journal of Naval Engineering*, Vol. 17, Book 1, (Jun., 1967), 7–15 (p. 10).

Fig. 6.1 Members of HMS/m *Dreadnought*'s ME Department in the machinery space

nuclear submarines would be expected to give a minimum return of five years' service and four years for other ratings. In view of the longer training HMS *Dreadnought*'s ratings would receive, it was decided that they should serve a minimum of seven years.[33] This was possibly to reflect the sea-training the men were to receive in US submarines; however, it can be argued that because HMS *Dreadnought*'s machinery was one-off purchased from the US, the Artificers and Mechanics selected to serve on her could be seen as specialised, having to operate and maintain US rather than UK machinery. It would cost the Admiralty time and money to re-train ratings to operate and maintain UK propulsion machinery that would be standard in the next class of submarines; therefore, it would make economic sense to keep ratings trained on HMS *Dreadnought*'s reactor plant and machinery specific to that submarine. The length of return of service

[33] DNP 2, NP184/2011, Meeting, Ratings conditions of service, 10 December 1958, p. 71.

certainly reflects the importance the Admiralty attached to nuclear training and their expectations of the men trained.

Practical Training and Jason

To support the work of the Department of Nuclear Science and Technology, a low-energy training reactor was installed at Greenwich. The building chosen to house the reactor was the historic King William Building, designed by Sir Christopher Wren and housing the famous "Painted Hall". It has been noted wryly that this was the only reactor building designed by Wren.[34] Although the Ministry of Works had been advised that the training reactor was completely safe, a draft letter to the Admiralty wanted assurances that there was no risk of radiological contamination to the historic building. Somewhat sardonically, the author of the letter wrote: 'Presumably you will seek the advice of a technically qualified scientist'.[35] The estimate for housing the original reactor in the basement and concreting the floor above was put at £16,560. There is a scarcity of information relating to the original reactor, only a reference to the existing sub-critical reactor which would be removed in preparation for installing a new training reactor.[36] By September 1961 it was apparent that the original reactor was to be replaced by a 10KW "Jason" reactor, specially designed for training nuclear engineers which had been operated by the Hawker-Siddeley Nuclear Company at their site at Langley, Buckinghamshire. Jason was shut down 19 June 1961 and dismantling was completed by October, the rebuilding work at Greenwich commenced 8 March 1962. Work on Jason was completed 24 September when mechanical commissioning started.[37] Jason went critical for the first time in November and continued as an instructional aid for students at Greenwich until shortly before the college closed in 1998.[38] With the expansion of nuclear engineers required to crew the new Polaris-equipped

[34] D. K. Brown and George Moore, *Rebuilding the Royal Navy: Warship design since 1945* (London, Chatham Publishing, 2003), p. 121 footnote.

[35] TNA, WORK 14/2691, Adaptions to King William Buildings for Department of Nuclear Science, 1953–64, Letter, to Chaff, 28 October 1959.

[36] TNA, WORK 14/2691, Letter, RNC/NS/1, 18 September 1961.

[37] TNA, POWE 74/83, Rebuilding of Jason at Royal Naval College, Greenwich, 1960–68, Report, p. 2.

[38] Our Naval Correspondent, 'Training Reactor Goes "Critical"', *The Times*, 8 November 1962, p. 6, Issue 55,543.

submarines (each Polaris submarine operated with two crews to maximise the submarines' time at sea), the Nuclear Science and Technology Department at Greenwich was required to increase the frequency of courses and the facilities to train these engineers. Up until 1963 there were two, sometimes three courses running concurrently at Greenwich, from the 1964–65 academic year there would be four and occasionally five courses run each year. The Department's classrooms and laboratories had cost £34,000 during 1959–60 and the proposed expansion which would involve moving the chemistry and physics laboratories included a similar amount of work. The estimate for this work, £68,500, shows a considerable financial commitment from the Admiralty in training its nuclear engineers (Fig. 6.2).[39]

Given the importance of this historic building and the nature of the research equipment, Jason, the low-energy experimental reactor, it is surprising that there was no debate on the subject in Parliament, although there is a reference tucked away in a Parliamentary Report of natural uranium fuel being supplied to Greenwich.[40] From this period, there is

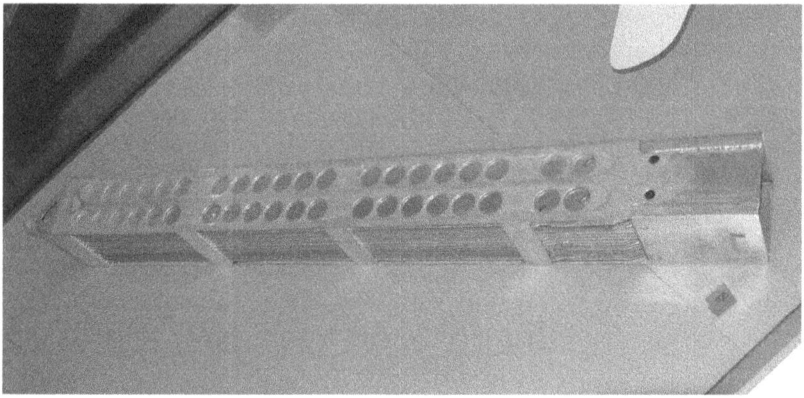

Fig. 6.2 Replica Jason Reactor fuel rod

[39] TNA, ADM 1/28870, RN College, Greenwich: major works items to provide nuclear propulsion training facilities, 1963–64, Register No. T.328/63, Minute Sheet No. 2, 30 September 1963.

[40] Parliamentary paper, United Kingdom Atomic Energy Authority. Seventh annual report for the period 1st April 1960–31st March 1961. 1960–61 (229), p. 51.

only one mention in *Hansard* of a reactor at Greenwich and this was during a debate on the Greenwich Hospital and Taverners' Foundation. The member for Islington North, G. W. Reynolds MP, on discussing the income the Foundation received in rental from the Royal Navy noted, that apart from the location and surroundings: 'The new tenants would have the use also of a wide range of useful assets, including [...] a nuclear reactor'.[41] If Parliament was not debating the siting of a reactor in a historic building, *The Times* was reporting that a training reactor was being sited at Greenwich, although, as one would expect of that period, the reporting is factual. No editorial comments or letters to the editor were found. It was hoped to research local newspapers from the period to see if there was any debate on Jason being sited at Greenwich but during 2018 the Greenwich Heritage Centre was in the process of being moved. Of the articles found on the training reactor, it was reported as part of the Navy's £2,500,000 expenditure on research and development a low-power training reactor was to be installed at Greenwich 27 February 1962, a larger report appeared when members of the press were invited into the Nuclear Department at Greenwich and shown around Jason; indeed, images of the control panel and the reactor were produced on the pictures page of that day's edition.[42] The reports in *The Times* dispel the myth that the Admiralty kept Jason's location a secret from the public; however, it does not explain the decision of Greenwich Council to declare itself a "Nuclear Free Zone" the same year, 1963.[43]

As recorded earlier, ratings passing the Long Course at HMS *Sultan* would go to Dounreay to attend a thirteen-week course with eight weeks being classroom based. During this period students would visit the submarine hull to find and trace the various systems and would sit two intermediate and one final exam.[44] In the late 1960s, the classroom element also included training on a simulator which closely resembled the manoeuvring room (Fig. 6.3), this early use of a simulator afforded a safe environment to allow students to become familiar with the various panels and

[41] Parliamentary debate, House of Commons sitting, 11 July 1963, Elizabeth II year 12, Columns 1385–1606, Fifth series, Volume 680.

[42] Our Naval Correspondent, 'Training Britain's Nuclear Sailors', *The Times*, 30 January 1963, p. 5, Issue 55612.

[43] 'Jason casts a cloud over naval college sale', https://www.independent.co.uk, [Accessed 18 February 2021].

[44] An in-depth description of the Dounreay Submarine Prototype is given in the Dounreay sub-chapter.

Fig. 6.3 Renown manoeuvring room. Courtesy the Friends of the Submarine Museum

conduct standard and emergency evolutions before keeping watches on the prototype reactor at Dounreay. Practical training was formed of three phases; general watch keeping in all the machinery spaces and on all panels in the manoeuvring room (where the reactor, electrical generation and propulsion systems were controlled) whilst the reactor was critical. The second phase of variable-power watch keeping covered pre-critical routines, taking the reactor critical, plant start-up, modes of propulsion, plant shut down and the connection/disconnection of electrical shore supplies. In the third and final phase, students would deal with various machinery drills which ranged from loss of main turbines to reactor scram and recovery.[45] Ratings on the shorter course would also visit the hull to find and

[45] Wilson, 'Nuclear Submarine Reactor Plant – Practical Training', pp. 11–12. Note: See technical definitions for the term Scram.

trace systems and sit exams before moving onto practical training on the reactor plant. The shorter three weeks practical training covered watch keeping in the machinery spaces and on the electrical and propulsion panels in the manoeuvring room. Apart from the courses described above, a four-week Health Physics course was also offered at Dounreay for medical officers and technicians as part of their Long Course prior to joining their submarines. This course gave students a period to practice reactor chemistry sampling techniques and other Health Physics procedures as well as experience of the equipment carried onboard and the operating conditions they would encounter.

Practical training at Dounreay was the first time that officers' and ratings' training was combined. Training in the classrooms and on the simulator was normal-day work; however, practical training was conducted on the prototype in a three-watch system. On successful completion of training at Dounreay all students faced a further two to three months' training onboard their respective submarines before their final examination to become fully qualified in their watch keeping position.[46] Continuous training is a part of service life as new and modified equipment is introduced or improved procedures are developed for operating equipment. In 1971, a new manoeuvring room simulator, "FASMAT", was opened in HMS *Neptune* at Faslane.[47] The nuclear engineers, as do all submarine officers and ratings, spend their careers between time on operational submarines and time in a shore billet. After a period ashore (typically two years), on moving to a new submarine, the officer/rating will once more re-familiarise themselves with the equipment and operating procedures before being examined to gain their watch keeping certificates. The Polaris submarines operated a two-crew system whereby the "on-crew" conducted maintenance and took the submarine to sea on exercises and patrol; the "off-crew" assisted with the maintenance period and once the submarine had sailed they would take leave and attend courses as required. After a period of approximately four months the submarine would return from its patrol and the crews would change over, this necessitated additional training of the off-crew nuclear watch keepers to bring them back to the required standard prior to going on-crew. The "FASMAT" trainer took some of the

[46] N. B. M. Clack, 'Practical Training of Nuclear Submarine Power Plant Operators', *Journal of Naval Engineering*, Vol. 19, Book 1, (Jun., 1970), 25–30 (p. 25).

[47] 'FASMAT Nuclear Submarine Simulator', *Journal of Naval Engineering*, Vol. 19, Book 3, (Jun., 1971), 482.

training load off Dounreay and was conveniently situated within the Clyde Submarine Base.

Dounreay Submarine Prototype (DS/MP)

As noted previously, Treasury approval for a shore-based submarine prototype reactor was received by the Admiralty in January 1956.[48] The next question was where to site the reactor; two possibilities were discussed, Windscale in Cumbria, and Dounreay on the north coast of Scotland. Dounreay was the site of the UKAEA's fast-breeder reactor situated on 360 acres of a disused Royal Navy Air Station; as such it was deemed the most suitable site for the prototype. In September, the Controller authorised the project to proceed, the First Lord added his approval a few days later commenting: '...that it was unfortunate but necessary to build at Dounreay'.[49] No doubt this a reference to its remote location and the difficulty of travelling to the area (especially in the 1960s when car ownership was not as widespread as it is these days). An Admiralty letter was sent to the Treasury advising them of the decision to site the prototype at Dounreay. The Admiralty informed the Treasury of the economic case in building the prototype on land adjacent to an establishment already operated by the UKAEA which was also large enough to allow expansion.[50] The programme, in the Autumn of 1956, envisaged the planning of the Dounreay site and facilities starting January 1957, building work beginning January 1958, with plant and machinery installation starting in March 1959. The installation of equipment was to be completed within a year and commissioning trials of the prototype begun by June 1960. HMS *Dreadnought*'s keel laying was planned for January 1960 and launch in June 1961, her commissioning trials were to begin in December. Dounreay had three main tasks, i) to operate and test submarine prototype machinery and advise on methods of improving the plant, ii) to advise on component design and maintenance procedures and iii) to train officers and ratings.[51]

It can be seen from the provisional 1956 programme, that it was originally planned for Dounreay to be operating approximately two years in

[48] TNA, T 225/1022, Letter, D.M.35/77/01, 6 January 1956.
[49] DNP 2, NP184/2011, Reference G.F.499/56, 3 September 1956, p. 23.
[50] TNA, T 225/1022, Letter, G.F.I./G.F.499/56, 18 September 1956.
[51] Clack, 'Practical Training of Nuclear Submarine Power Plant Operators', p. 26.

advance of HMS *Dreadnought*, so any problems encountered with the reactor design or its operation could be resolved or re-engineered prior to fitting into the submarine. The Admiralty would share UKAEA facilities on the Dounreay site, with the UKAEA providing transport infrastructure, workshops, laboratory and health services, housing and security police. There would be no capital costs to the Admiralty for these services but there would be an annual charge. The UKAEA was willing to act as the Admiralty's agents and design all the building and civil engineering work using their own contractors. This made economic sense given the remoteness of Dounreay and the shortage of accommodation in the area for contractors, the nearest town of significant size being Thurso ten miles along the coast to the east. The Admiralty expected the cost of establishing Dounreay to be around £1,775,000, the total research and development cost for the submarine programme was estimated at £12,050,000 between 1956 and 1961 (excluding approximately £7 M for the submarine).[52]

The Treasury, with an eye on the possibility of reducing the costs of the programme, had asked the Admiralty what effect the 1956 amendment to the 1955 Agreement would have on the programme. The 1956 amendment covered the exchange atomic information for civil uses which would also allow the release of information on naval propulsion reactors.[53] The Admiralty stated their case strongly advising that there was no clear indication of when the information would be released; it was their experience (and that of the UKAEA) that the US would not offer information if there was to be no reciprocal original information to be exchanged. The uniqueness of the materials used and the problems of producing them with the same chemical and radiological qualities as those in the US meant that the materials produced in the UK also had to be proven. The Admiralty argued that they had to continue with the project as quickly as possible so as not to fall further behind noting that: '...although any U.S. information which we obtain may well result in a slight reduction in our own work, it cannot replace it'.[54] Although Treasury authority for the nuclear submarine project had been received in January with an agreed expenditure for 1956/57 of £300,000, the Admiralty wrote to the Treasury to sanction a sum of

[52] TNA, T 225/1022, Letter, G.F.I./G.F.499/56, 18 September 1956.

[53] See Chap. 5 and the problems surrounding Eisenhower's minority administration in releasing this information.

[54] TNA, T 225/1022, Letter, G.F.I./G.F.499/56, 18 September 1956.

£588,000 during that year. Of this money, £238,000 was to be spent at Harwell and other components of the UKAEA and £350,000 would be spent on the development of the prototype machinery at Dounreay. The main propulsion machinery amounted to £84,000, the reactor vessel £43,000 and the research rigs and heat transfer experiments £225,000.[55]

The main contractor for the Dounreay installation was Vickers-Armstrongs (Engineering) who had responsibility for the design of the whole machinery. Rolls-Royce was responsible for the reactor design and associated equipment comprising the core, fuel elements, emergency cooling system, thermal shield and fuel handling equipment. Foster Wheelers had the task of designing most of the equipment in the reactor compartment, such as the reactor pressure vessel, steam generators, main circulating pumps and the primary piping and valves.[56] When completed, the Dounreay naval establishment was comprised of an administrative block containing offices, an accommodation centre (Scapa House), classrooms and the main building which housed the reactor prototype, workshops and auxiliary machinery. The reactor plant had to be designed to fit inside the envelope of the submarine's pressure hull and to be installed during the submarine's build. Therefore, the prototype was to be built into a structure that would simulate a section of the submarine pressure hull. The reactor compartment, housing the primary system, was a twenty-eight-feet cylinder, thirty-four feet long enclosed by ribbed bulkheads.[57] The submarine propulsion machinery was housed within the building and laid out as it would be on a submarine but without being fully enclosed by a pressure hull. A pumping station was built on the foreshore with enough capacity to provide cooling water to the prototype 250 yards away. A water tank of 2000 tons capacity was built around the reactor compartment section, allowing emergency cooling by means of convection should the sea-water circulating system fail.[58] It is evident that the Admiralty and its contractors were taking measures so that the prototype plant and machinery would be operated in conditions as realistically as possible so that it ensured any operating data or system function would be replicated on the submarine.

[55] TNA, T 225/1022, Letter, G.F.I./C.P.54974/56, 9 November 1956.
[56] Edwards, 'Joint Panel on Nuclear Marine Propulsion…', p. 2.
[57] TNA, AB 8/791, Note, Anon. 7 March 1957.
[58] TNA, AB 8/791, Letter, 330/70/1, 13 June 1957.

Dounreay's Future Questioned

During this period there was concern within the Admiralty that the Prime Minister, Chancellor and the Defence Minister were not fully committed to the nuclear submarine project. The Prime Minister wrote to the First Lord, the Earl of Selkirk, asking for a brief history of the atomic submarine; when was it first authorised, how far had research and development progressed, to what operational use would the submarine be put and its future role in the Navy. Macmillan noted: 'I am afraid that I cannot recollect how this undertaking began'.[59] It is quite apparent that the Admiralty viewed this Minute with a measure of distrust as to its motives. In the Minute, the Admiralty highlight Macmillan's referral to the submarine in the singular which reflected his view of the development as an isolated project. The Admiralty took the position to offer only the information requested and a draft memorandum was discussed prior to releasing it.[60] Selkirk replied with a robust argument on the operational use of nuclear-powered submarines (referred to in the plural), their stealth and their endurance capabilities. Selkirk also advised that the Dounreay site was nearly complete and important modifications had been included in the design of the prototype plant from information received from the US.[61] The modifications Selkirk refers to would have included the decision to proceed with uranium/zirconium fuel elements.

Shortly after Selkirk's memorandum, the USS *Nautilus* paid a visit to the UK and held a VIP day at sea, as noted in Chap. 4, Sandys was amongst those who went to sea. The USS *Nautilus*' visit coincided with the Royal Navy exercise "Rum Tub", 14–19 October, in which surface and submarine elements of the Royal Navy were pitched against the USS *Nautilus*. No doubt Sandys would have debriefed Macmillan on his return to London; he was recorded as saying in the *Daily Telegraph* that: '...this submarine represents in the sphere of naval warfare a revolutionary advance as great a change from sail to steam'.[62] Mountbatten certainly thought that Sandys was a keen supporter of the project prior to his trip in the USS *Nautilus* which he was looking forward to; with Sir Frederick Brundrett also an advocate of HMS *Dreadnought*, Mountbatten felt

[59] TNA, PREM 11/2635, Minute, H397/57, 7 August, 1957.

[60] TNA, ADM 116/6204, Atomic reactors: exchange of information with the USA, 1956–57, Register No. M.601/11/57, Minute Sheet No. 1, 22 August 1957.

[61] TNA, CAB 21/4852, Memorandum, Selkirk to Macmillan, 12 September 1957.

[62] Broadlands Archive, MB1/I300, Quarterly Newsletter, 1 November 1957, p. 10.

confident that it would be extremely difficult for the Government to cancel the project.[63] Indeed, as noted in Chap. 3, from the moment of Sandys' appointment to the Ministry of Defence, Mountbatten had endeavoured to create a good working relationship with him using his influence and contacts to that affect. There can be no doubt that these occasions did have a positive impact on Sandys' relationship with Mountbatten.

By January 1958, the site foundations had been laid, roads constructed and building had commenced; machinery installation would start in May. The financial estimates looked good for Dounreay, the total for the site works and machinery amounted to £1.83 M, an increase on the September 1956 estimate of just £55 K. This would appear to indicate that the Admiralty had a firm grip on the financial side of the prototype.[64] However, early 1958 brought new challenges resulting from Rickover's offer to sell to the UK a complete submarine propulsion unit; would Dounreay now be necessary? Plowden's view was that all work specifically directed at the development of the submarine reactor should cease immediately without waiting for Rickover's proposals to be completed. Plowden had also been advised that Rolls-Royce had already removed their design teams from the Dreadnought project. It should be remembered that, apart from the heavy workload connected to the expanding civil nuclear power station generating programme, there was the fire at the Windscale plant in October 1957 that required the UKAEA's urgent attention. Therefore, it is entirely plausible that the UKAEA was under pressure to reduce some of its commitments to the Admiralty's work, to which the Admiralty would have been sympathetic. Plowden and Lang agreed to cease work on the submarine reactor and to discuss the current programme at Harwell.[65] It had also been decided separately that all work already committed at Dounreay was to continue; however, the administration building, canteen and houses would be deferred.

A meeting was held to discuss the future of the programme, in seeking Cabinet approval to purchase *Skipjack* machinery the Admiralty had advised that it would result in a reduction of the project's research and development effort. No doubt the Treasury would be looking for substantial savings, the Admiralty was wise to this possibility and Lang advised: 'It

[63] Broadlands Archive, MB1/I397, Sea Lords' Meeting 8 October 1957.

[64] TNA, DEFE 7/2055, Memorandum, M.618/5/57, 14 January 1958, appendix II.

[65] TNA, POWE 14/1157, United Kingdom Atomic Energy Authority: Admiralty submarine programme, 1958, Letter, CH(58)27, 25 February 1958.

would be prudent to give some prominence to research which was being discontinued rather than that which we hoped to continue'.[66] It was recorded that work on the facilities at Dounreay were progressing and that it would be imprudent to stop work without the US responding favourably to the request for machinery. Dounreay was also thought to be essential for any future submarine development in the UK; therefore, it was decided to complete Dounreay as presently designed (with British machinery) and to include the primary circuit as this had already been ordered and was in manufacture. The Admiralty wrote to the Treasury with an outline of the revised programme, reiterating their intention not to "Anglicise" the S5W reactor plant but rather develop the Dounreay plant with the lessons learnt from the US plant. Thus, in justifying the continuation of research and development at Dounreay, the Admiralty would make full use of the efforts already expended in developing the site. Importantly for the Treasury, the Admiralty highlighted the £500,000 savings in research and development during 1958/59 and about £700,000 in each subsequent year leading up to development and fabrication of their own S5W fuel element, '...the emphasis in work that is continuing is very much on development rather than research'.[67] This line is rather telling as it indicates that the Admiralty was confident that enough research had been done and that development would progress from the successful incorporation of US knowledge and the nuclear engineering skills the UK was mastering.

Rolls-Royce had carried out important design studies for the Admiralty of the Dounreay reactor pressure vessel and advised that it would accommodate the S5W core.[68] Rolls-Royce's report also recommended the use of stainless steel for the primary circuit and other changes reflecting the advice of Rickover, but this is fascinating.[69] Undoubtedly, a generic engineering problem tackled by two separate teams working in isolation from each other could, theoretically, result in the same solution, especially if the design requirement is to realise similar objectives. The US Navy was designing a 15,000 SHP power plant to propel a single-shaft, 3500-ton submerged submarine and the Royal Navy was designing a 20,000 SHP

[66] TNA, DEFE 69/749, Note of second meeting to discuss the nuclear submarine project. *The date on this note, 21 March 1957, suggests it is a typographical error and should be 1958.

[67] TNA, ADM 116/6411, Letter, G.F.247/58, 10 April 1958.

[68] Hill, 'Admiral Hyman G. Rickover USN...', p. 10.

[69] Lambert, *Rolls-Royce: the Nuclear Power Connection*, p. 42.

power plant to propel a single-shaft, 4000-ton submerged submarine. Working separately, both teams managed to produce similar designs with little to separate them, as is intimated from the references to Rolls-Royce's design study above. It is intriguing, therefore, to reflect on Goodman's assertions of atomic intelligence exchanges continuing in the face of the US McMahon Act and, indeed, Baylis' claims of a "transatlantic advocacy coalition" of like-minded nuclear scientists and defence officials discussed earlier.[70] Whilst this may be true of other areas where limited nuclear information was exchanged, it is evident from all the files researched that there appears to be no record of any such clandestine cooperation with respect to the nuclear propulsion programme, and the designs of the S5W and PWR 1 must, therefore, be seen as a result of the endeavours of two separate teams with a comparable design philosophy.

A re-phased development programme was set out in August detailing what work was to continue and what was to cease. The design and production of the machinery would continue at Dounreay, with the exception of the reactor core and control mechanisms which would only proceed once experience had been gained of the US design.[71] In fact, it was reported that the British plant planned for HMS *Dreadnought* and the S5W plant shared almost identical operating and performance characteristics; therefore, the Admiralty was confident of the Dounreay plant operating with a S5W core. Indeed, one of the reasons behind the Admiralty's decision not to "Anglicise" the S5W was they thought that: '...the British design is considerably in advance of the American in certain respects, particularly those of silencing and shock'.[72] The Admiralty's policy at Dounreay could now be summarised thus: i) the design principles governing Dounreay's machinery were mainly based on US development, so lessening the research and development cost; ii) manufacture of the machinery was well advanced and, therefore, would not be an exact replica of the S5W plant with the exception of the core and control mechanisms as advised previously; and iii) future submarine machinery would be all British design based predominantly on the Dounreay plant and further developments from it, thereby reducing UK reliance on US machinery with the possible

[70] For further details on Anglo-US sharing of atomic knowledge see Goodman, 'A Little Help from my Friends:' and Baylis, 'The 1958 Anglo-American Mutual Defence Agreement:'.

[71] TNA, ADM 1/27375, Note, Controller to Baker, 6 August 1958.

[72] TNA, ADM 1/27750, Letter, G.F.247/58, 10 April 1958.

development of higher power rates.[73] Professor Edwards noted in his paper on the initial problems of the pressurised water reactor design that it was a great disappointment to all that the first reactor to be fitted into a Royal Navy submarine would not be British, but it was satisfying to know that much of the work of the team was being embodied in the machinery to be installed at Dounreay. 'The work carried out to the end of 1958 was therefore of considerable value'.[74] Considering that in some respects the British design was in advance of the S5W plant and that the Dounreay plant was fully utilised apart from the core and control rod mechanisms, it is perhaps understandable that some in the Admiralty team felt great disappointment at the decision to buy the S5W plant.

Edwards, however, noted in the same paper that whilst the decision to purchase the S5W plant gave the UKAEA some relief and accelerated the completion of HMS *Dreadnought*, it was at the expense of the prototype which inevitably ran into delays. Vice Admiral Sir Ted Horlick perhaps best summed up the delays incurred at Dounreay. Despite the quality and dedication of the teams involved it was not the reactor theory but engineering in its broadest sense which provided the delays, the requirement to handle and fabricate new materials, the design of new welding techniques and their acceptance standards had to be acquired: '…working to tighter tolerances, specifications, and requirements for quality assurance together with very high standards of cleanliness; all this had to be learnt the hard way—by experience'.[75] Most notably, however, whereas great attention to detail was given to the primary system, little attention was given to the secondary plant as it was regarded as established technology. In effect, apart from the quality assurance and control imposed on the manufacture of the reactor plant and primary components which were designed to fit into the envelope of the submarine's pressure hull, the secondary machinery: '…was rather taken for granted'.[76] Taking for granted the secondary systems and not giving them the due diligence that was given the primary system was not only a UK oversight, there was possibly the same mentality in the US that ultimately led to the loss of the USS *Thresher* (a nuclear-powered submarine fitted with a S5W reactor) on 10 April 1963. It should also be noted that many of the machinery elements

[73] DNP 2, NP184/2011, Admiralty policy at Dounreay, p. 57.
[74] Edwards, 'Joint Panel on Nuclear Marine Propulsion…', p. 1.
[75] Horlick, 'Submarine Propulsion in the Royal Navy', p. 7.
[76] Horlick, 'Submarine Propulsion in the Royal Navy', p. 8.

installed at Dounreay had never existed and there was no experience of operating such a plant. The lack of experience and the difficulty of access would have made operation and maintenance of the machinery problematic; this led to delays installing the machinery, which required significant redesign and modification.

Dounreay also suffered from management problems during its build programme. Vickers-Armstrongs had a separate technical organisation for electrical work which meant that there was no person in overall control of the project; this was totally at odds with Rickover's much quoted concept of responsibility which he made at the enquiry into the loss of the USS *Thresher*.[77] The Admiralty insisted that one man was put in overall control, but he was deemed unsuitable by the Admiralty although they could not immediately criticise the appointment. A new man was eventually appointed, and a complete reassessment of the project was conducted, which, by the end of 1961, was completed and a new system put in place.[78] The company had been lax in their time and cost estimating, and financial control; they also suffered from their fair share of strikes at the site. 'This was a case where a unique project had to be grafted into an organisation notable for its conservative and traditional attitudes'.[79] The Admiralty, however, conceded that Vickers-Armstrongs was the firm best equipped technically to undertake the project which was now considered satisfactory. The Admiralty also had criticism for the UKAEA which they thought had been uncooperative with their estimating. Admiralty justification for continuing with Dounreay had been accepted but these engineering problems were not the only delays, there were still material concerns to overcome.

One other problem to overcome was obtaining zirconium and hafnium of a purity that was not available in the UK. Although this material was included in the purchase of the complete machinery for HMS *Dreadnought*, there was no material available for the Dounreay plant. A contract was placed, through Rolls-Royce, with I.C.I. to purchase from the US 1700 pounds of hafnium crystal bar and the forty-two tons of zirconium sponge associated with its production. The estimated cost was £248,500; even

[77] For a transcript of Rickover's concept of responsibility see Hill, 'Admiral Hyman G. Rickover USN…', p. 15.

[78] TNA, T 225/2206, Admiralty development of nuclear submarine prototype at Dounreay Scotland, 1960–64, Minute, 2-D.M.209/32/01, 4 December 1962.

[79] TNA, T 225/2206, Minute, 2-D.M.209/32/01, 4 December 1962.

then the Admiralty was not satisfied that the material would be delivered in time to meet Dounreay's programme.[80] A UK company, Magnesium Elektron Limited, had a pilot plant for producing hafnium in the UK; the Admiralty approached the company to provide insurance against failure of supplies from the US. If the company managed to fulfil the order the Admiralty also saw it as a means of becoming self-sufficient. Resulting from the purchase of the S5W plant and the use of hafnium for the control rods, the Admiralty decided not to ask the UKAEA to investigate other possible control materials such as, silver, cadmium and indium, which would result in a £212,000 savings over four years. This suggests a certain conviction in the Admiralty that they would receive the hafnium at the purity they required, it was also an acknowledgement that hafnium was a far superior neutron absorber than the other materials, as discussed in Chap. 3. It was certainly a means of illustrating to the Treasury that the Admiralty was taking measures to curtail their research expenditure.

A Very Serious Snag

Due to delays in 1958/1959 it had been recognised that it would have been untimely to determine the arrangements for running the Dounreay prototype; however, by early 1960 the Admiralty had concluded that the time was right to discuss the matter with the firms involved. Once the prototype had been commissioned, the Admiralty envisaged that a Royal Navy Commanding Officer with a small team would be responsible to the Admiralty for the running of the establishment. However, it was the Admiralty's intention to select a contractor to maintain and operate the prototype plant and machinery and that the contractor would appoint a Chief Test Engineer responsible to the Commanding Officer. The Admiralty required the Chief Test Engineer to be: '…a person of the highest calibre, possessing the necessary technical and administrative ability […]. He must also meet with the approval of the A.E.A'.[81] Building continued during 1960/1961 and safety reports were assessed at each milestone. In a letter to the Treasury requesting the release of funds it was advised that: '…there are no known serious snags ahead on either the nuclear or conventional machinery side'.[82] There was, however, a caveat

[80] TNA, T 225/1193, Letter, GFI/DRPP84/59, 12 June 1959.
[81] TNA, AB 6/2308, Letter, C.P.52769/59, 29 March 1960.
[82] TNA, T 225/2206, Letter G.F.I./S. 10219/61, 21 May 1962.

advising that Dounreay was a development project and subsequent testing could still be a source of trouble. As things turned out this was a prescient warning, the Admiralty was soon to run into a very serious snag. As the Admiralty's main contractor responsible for Dounreay, Vickers-Armstrongs' site manager Noel Davies, was selected as Chief Test Engineer in 1962. In 1964, Rolls-Royce and Associates' site manager, Hugh Eaglesfield, was appointed Chief Test Engineer. It was during this important period that responsibility for Dounreay would pass from Vickers-Armstrongs to Rolls-Royce and Associates as the Admiralty's operating contractor. Farley-Sutton noted, both men would be at the heart of the eventual success of the prototype project.[83]

The UKAEA's participation in the prototype included assisting in the commissioning, calibration and operation of the plant, advising on the subsequent operation of the plant and its use as a test facility, examination of irradiated fuel elements and limited examination of irradiated components. One further important area of responsibility was obtaining chemical and metallurgical information on the corrosion problems of the ferritic circuit.[84] This last task was crucial, the prototype primary system was identical to HMS *Dreadnought*'s; the main difference was the core being manufactured under license in the UK by Rolls-Royce. However, one other important difference was that the primary circuit of the prototype was manufactured in part from chrome-molybdenum and a nickel-based heat-resistant alloy known as Inconel; in contrast, HMS *Dreadnought*'s primary circuit and reactor pressure vessel were manufactured from stainless steel. The decision to use chrome-molybdenum and Inconel rather than austenitic stainless steel resulted from the lack of data on stress corrosion, irradiation damage, hydrogen embrittlement and other related issues. Despite a growing awareness of stainless steel metallurgy, its susceptibility to chloride stress corrosion outweighed low-alloy steel's relatively higher corrosion rate and slight susceptibility to hydrogen embrittlement.[85] There is little doubt that cost also played a part in this decision. Edwards notes that later the US began to give attention to low-alloy steels to reduce capital costs, Rickover was interested in the use of chrome-molybdenum but advised against using untried materials in a submarine. It was also noted that the Americans made no criticism of employing Inconel which both

[83] Private papers, Captain Colin Farley-Sutton RN Ret'd.
[84] DNP 6, CP60801, Memorandum, 687(R) in covering Letter, Fletcher 27 July 1959.
[85] Edwards, 'Joint Panel on Nuclear Marine Propulsion…', p. 20.

navies regarded as an established material and one which Rickover was known to favour.

The auxiliary primary circuit consists of small bore pipework and although making welds in non-ferrous metals is difficult, the use of Inconel was working satisfactorily. The welds were: '...properly and independently tested, had been radiographed and had withstood deliberate over-pressure on many occasions'.[86] However, in September during pre-criticality testing when the primary circuit was hot and pressurised, a number of leaks appeared in the small bore pipework and within weeks the number of leaks in the Inconel welds had risen to twelve. Although a major investigation was implemented to discover the cause of this failure, it was apparent that: 'With one thousand six hundred and fifty-eight nickel alloy welds in the DSMP primary plant and no clear solution in sight, the decision was taken early in 1964 to renew all nickel alloy piping and fittings in chromium-molybdenum low alloy steel'.[87] This serious problem had repercussions for submarine building; Horlick noted that replacement in stainless steel for the prototype had been considered, but the second submarine had more pressing needs on the limited supplies. HMS *Valiant*'s primary circuit had already been fitted and was deemed to be suspect so would require removing. The third submarine, HMS *Warspite*, and the Polaris submarines were all to be fitted with Inconel primary circuits; the decision was taken for these submarines to revert to stainless steel circuits. Some redesign work was anticipated, and it was advised that manufacture and installation of the new stainless steel circuit would cause a delay of fourteen months in HMS *Valiant*'s completion. The only recourse to mitigate the delay was to purchase supplies from the US, an approach that was reliant on the US having stock available and a willingness to release material to alleviate the delay anticipated. It was estimated that changing from Inconel to stainless steel would add £500,000 to the £19.6 M for HMS *Valiant* and between £100,000 and £500,000 for HMS *Warspite*; there would also be a small unspecified additional cost to the Polaris submarines as these were not yet laid down.[88] The Inconel problem at Dounreay more than justified the prototype's requirement and its worth to the fledgling nuclear fleet as the defect could be rectified or re-engineered before fitting into a submarine where rectification would have been more problematic, and expensive.

[86] TNA, T 225/2206, Letter, MacMahon to Bell, 25 November 1963.
[87] Horlick, 'Submarine Propulsion in the Royal Navy', p. 8.
[88] TNA, T 225/2206, Letter, Bell to MacMahon, 27 November 1963.

This problem was being discussed between Admiralty and Treasury officials in mid-November, President Kennedy was assassinated 22 November 1963, and out of respect it was decided to delay approaching Rickover. The first attempt to purchase stainless steel did not bode well; Theodore Rockwell (Rickover's Technical Director) advised that no progress could be made until the material requested corresponded more closely to two sets.[89] These sets were required for HMS *Valiant* and HMS *Warspite*; however, it can be appreciated that the two British submarines were based on the Dounreay prototype and that there would be subtle differences between what the Admiralty required to fit-out the submarines and the original sets manufactured for HMS *Dreadnought* (*Skipjack*). In the New Year, Sir Solly Zuckerman, the Chief Scientific Adviser, wrote to Rickover to indicate these differences and to allay any concerns he may have that the Admiralty was over-bidding their requirements. The request was for the minimum amount of material to fit-out the two submarines with an allowance of ten to fifteen percent for spares.[90] An export licence was soon granted enabling the contract to be placed and, as if to highlight the limitation of the 1958 Agreement, the draft application included words to the effect that with Rickover's authorisation the UK could buy improved Dreadnought valves. Rockwell objected advising that whilst he would not exclude the possibility if absolutely necessary: '…to attempt to stray beyond Dreadnought type valves in the application would create difficulties'.[91] Although there were many impediments in getting the Dounreay prototype commissioned and in building the new submarines, through the early identification and resolution of this material issue alone the prototype had proved its worth from the offset.

Finally, at 03:10 7 January 1965, after nine years of delay and frustration the first British submarine reactor went critical for the first time. Farley-Sutton described it as: '…a marvellous, and chastening, and surprisingly calm moment'.[92]

[89] TNA, DEFE 19/243, Arrangements for re-fuelling HMS Dreadnought, 1963–66, Letter, ACO(W)550/02, 30 December 1963.

[90] TNA, DEFE 19/243, Letter, SZ/18/64, 2 January 1964.

[91] TNA, DEFE 19/243, Letter, ACO(W)550/02, 10 February 1964.

[92] Private papers, Captain Colin Farley-Sutton RN Ret'd.

HMS/M *Valiant*

The first submarine to be commissioned with the British reactor fitted with Core A was HMS *Valiant*, commissioned into the Royal Navy 18 July 1966. During 1960, a comparative study had been conducted of the relative merits of the S5W plant with that intended for the Dounreay plant; this resulted in HMS *Valiant*'s plant differing from the Dounreay plant; however, some of the recommendations were not adopted until the *Swiftsure* class later in the decade. The major changes implemented were as follows: i) the fabrication of the primary circuit and major components in stainless steel, ii) the use of a low-alloy steel reactor pressure vessel lined with stainless steel, iii) simplification of the secondary plant to reduce the number of valves and fittings, iv) provision of a circulating water cross-connection between the main engine and turbo-generator condensers; v) design for improved access and vi) a reduction of the number of remotely operated valves in the reactor compartment.[93] To assess the capability of the submarine and her equipment during a prolonged period at sea, HMS *Valiant* was tasked to undertake a dived transit to Singapore. The voyage would see all machinery operating in various climatic conditions, from the cold winter waters of the North Atlantic to the warmer climes of the Indian Ocean, enabling a full assessment to be made of the effect of high sea water temperatures on machinery performance.

At 11:30 30 January 1967, HMS *Valiant* slipped from the depot ship, HMS *Maidstone*, at Faslane and made passage down the Clyde; she dived the following day and began her journey south during which time the crew conducted machinery evolutions and training drills. Travelling deep and at high speed, she passed Madeira a few days later, by Valentine's Day she rounded Cape Agulhas, the most southerly point of South Africa, and entered the Indian Ocean. HMS *Valiant* spent 19–24 February at Mauritius before continuing to Singapore where she went alongside another submarine depot ship, HMS *Forth*, 6 March. HMS *Valiant* conducted exercises with the Far East Fleet and the Seventh Submarine Squadron to give them experience of working with, and in opposition to, a nuclear submarine. Unfortunately, she had to return to Singapore early due to a gearbox defect. HMS *Valiant* began her return journey on the evening of 29 March. Once more, HMS *Valiant* dived and began a deep fast transit returning to Faslane 25 April, she had steamed 26,000 miles at

[93] Horlick, 'Submarine Propulsion in the Royal Navy', p. 9.

an average speed of 20 knots; she had missed just one commitment, the exercise off Singapore.[94] During this prolonged period at sea and unsupported it was inevitable that there would be defects with the machinery and other equipment; the reactor plant, however, performed entirely to expectations.

As Horlick had noted, not as much attention had been given to the secondary machinery as had been given to the reactor plant. The secondary machinery, too, was new to the fleet and nearly all the secondary propulsion defects: '...pointed directly at the questionable design and inadequate quality control of many items of the conventional propulsion and ships service equipment'.[95] It was believed that the greatest lesson learnt from the voyage was that conventional machinery should be built to the same high standards and quality as the nuclear plant. Indeed, as noted above, the loss of the USS *Thresher* in 1963 was attributed to a failure of a brazing joint in her sea water system (a secondary system). The loss of the USS *Thresher* caused the US Navy to modify their reactor operating procedures; this information was passed to the Admiralty, so they too would benefit from the lessons learnt; however, once the Rolls-Royce/Westinghouse contract expired, the Royal Navy would no longer profit from US reactor operating experience.[96] Following the loss of the USS *Thresher*, the US Navy instigated a quality assurance programme called SUBSAFE, which included more vigorous quality control and testing of secondary machinery during manufacture.

The Royal Navy introduced a similar system known as First Level, which deals with aspects of reactor safety and containment, equipment and systems that are essential to the watertight integrity of a submarine, and personnel safety. In the Royal Navy system, all material employed in First Level equipment is subject to a Quality Assurance Statement of Requirement (QASOR) which lays down the material specifications and tests which must be adhered to. Any deviation from any of the stated requirements would have to be agreed between the MoD and supplier and a concession certificate raised and placed with the necessary documentation in a quality assurance document pack which should remain available for the whole period the equipment is in service. The quality assurance

[94] Private papers, Captain John Jacobsen RN Ret'd.

[95] Private papers, Captain John Jacobsen RN Ret'd.

[96] TNA, DEFE 72/45, Register No. S.10583/63, Minute Sheet 1, paragraph 3a, 20 August 1963.

documentation is inspected when the equipment is receipted into MoD stores and again by ship's staff when the equipment has been issued for fitting into a submarine. It is this system that ultimately controls the materials used in submarine equipment. When HMS *Valiant* was ordered in 1960 the Polaris Sales Agreement (PSA) was still three years hence and it was certainly not on the political horizon. However, following the PSA, the reliability of the primary and secondary plant fitted in HMS *Valiant* took on a new importance as it would also be the plant fitted to the new Polaris submarines. Therefore, *Valiant*'s return voyage to Singapore was an important assessment of the plant's capability. The quality and reliability of the plant and of the secondary machinery was paramount if the credibility of moving the deterrence to sea was to be proven and maintained.

This chapter has examined the great strides the Admiralty made in getting officers and men educated and trained in the operation of pressurised water reactors. Although initial training for officers was conducted through somewhat inadequate courses at university, Royal Naval courses were soon established at Greenwich and HMS *Sultan*, Gosport, based mainly on the US Navy's curriculum. Classrooms, workshops and eventually simulators were built to accommodate the training requirements of the new nuclear fleet, and a system of examinations founded. With Dounreay still under construction and the prospect of HMS *Dreadnought* completing two years ahead of schedule with the purchase of the S5W plant, there was certainly pressure to get sufficient men academically trained to take advantage of practical courses offered with the US Navy. Pre-training in the UK was essential if Royal Navy ratings were not to be at a disadvantage to benefit from the same courses as US Navy ratings at Groton and it is apparent that effort had been made to ensure this did not happen.[97]

The Treasury had seen the purchase of the S5W plant as a means of saving money on research and development to which the Admiralty had cut a number of experiments and associated research once the contract was signed. The Treasury had also hoped for savings with the Dounreay plant; however, this overlooked the fact that all the conventional propulsion machinery and associated auxiliary systems were British designed and built, and that the majority of the reactor plant was also of British design. The Admiralty had already warned the Treasury that attempts to "Anglicise" the S5W plant would involve a tremendous amount of

[97] TNA, ADM 1/27088, Minute, RANP.14/11/58, 20 November 1958.

redesign and cause serious delays.[98] The Admiralty's reasons to continue with the British design of PWR 1 at Dounreay using the US core and control system was, in the end, justified as it allowed cores of greater power and greater endurance to be designed and back-fitted to submarines already in commission with the Royal Navy.

[98] TNA, ADM 116/6411, Letter, G.F.247/58, 10 April 1958.

CHAPTER 7

Future Developments

The Dounreay prototype's objectives were to formulate the starting-up procedure and assessment of safety and operational problems, to evaluate problems relating to nuclear plant maintenance and 'To provide a test platform for newly developed auxiliary arrangements and materials'.[1] Importantly, for future development, Dounreay was to provide comprehensive trials data associated with running a pressurised water reactor with varying loads which would allow correlation to be made between nuclear physics and heat transfer theories and the results obtained in practice. The data could then be evaluated, and the findings would allow improvements to be made in shielding design, efficiency in the core and increased output. Long-term planning for an improved core started before the initial Dounreay core, based on the S5W core, went critical.

As previously discussed, the reactor and primary systems at Dounreay were of a British design; the core and control rod mechanism were of American design and were incorporated into the Dounreay plant. This chapter will investigate the consolidation of the knowledge transferred from the US and the subsequent improvements in UK core designs which resulted in higher power ratings and longer endurance. The secondary

[1] TNA, AB 8/791, Minute of meeting 19 September 1958.

G. M. Jones, *The Development of Nuclear Propulsion in the Royal Navy, 1946–1975*, Security, Conflict and Cooperation in the Contemporary World,
https://doi.org/10.1007/978-3-031-05129-6_7

propulsion machinery at Dounreay was designed for 20,000SHP and would allow for an increased power output without having to build a new prototype. With future core developments, the UKAEA would still be required to assist in some of the programme; however, Rolls-Royce, now with their own nuclear experience, negotiated for references to design rights and patents not to be included in the new development contract.[2] Rolls-Royce wanted the opportunity to commercially use their inventions and designs emerging from their development work, a view that was unacceptable to the Admiralty and the UKAEA. This chapter will also look at other types of reactor designs. Experience with the S5W plant gave the Admiralty confidence to investigate other types of nuclear reactors for application in submarines although ultimately, for a number of reasons, they were not adopted by the Royal Navy.

Refuelling Preparations

On his appointment as Rear Admiral Nuclear Propulsion in February 1957, G. A. M. Wilson had a number of responsibilities. The memorandum announcing his appointment noted that amongst his duties was a responsibility to advise the Director of Dockyards and other Admiralty departments concerned, on the requirements for fuel handling and replenishment, the requirements for refitting nuclear-powered submarines and the training of dockyard personnel involved in the refuelling process.[3] Wilson dealt with the preparation of dockyards in a paper, D.M.3826/59 dated 19 May 1959, in which it was advised that the choice of dockyard should take into account the possible hazards attendant on the refuelling procedure. The UKAEA carried out a classification of UK ports on behalf of the Admiralty; the ports were graded A, representing a minimum hazard, through to G, which represented a place where a nuclear accident was deemed to have the most severe repercussions. Of the four major Royal Dockyards, the most favourable was Rosyth as it represented the most remote location, situated on the north bank of the Firth of Forth. The other dockyards, Chatham, Portsmouth and Devonport, were situated in areas of high population densities.[4] To decide on which dockyard to focus, the Admiralty appointed Captain D. P. Sparham to the Dockyard Division

[2] DNP 6, CP60801, Letter, Hedger to Barman, 5 April 1960.
[3] DNP 1, ALO 02617/97–000378, Memorandum, Ref 36/57, 15 February 1957.
[4] DNP 2, NP184/2011, p.81.

for six months to assess the effort required to prepare dockyards for refitting nuclear-powered submarines. Although the Royal Navy had yet to commission their first nuclear-powered submarine, action was already in hand to ensure dockyard facilities would be prepared to receive their first docking. HMS *Dreadnought*'s staff requirements envisaged the submarine docking every twelve months, for maintenance, and docking every two years for refit; the second refit was to be extended by twelve months to allow for survey of underwater equipment.[5] It can be appreciated that as the nuclear submarine fleet expanded, and especially with the procurement of Polaris, there was the additional strain of maintaining a continuous deterrent in keeping one of the Polaris submarines at sea at all times. Rosyth, being the smallest of the Royal Dockyards with just three docks, would never be able to cope with the demands of having two nuclear submarines in dockyard hands. A second site would eventually have to be chosen to service the nuclear fleet's requirements, of the three southern dockyards, Chatham was chosen as the second dockyard to refit nuclear-powered submarines. Work began on building nuclear refitting infrastructure in 1968 and the first submarine, HMS *Valiant*, entered into refit in 1970. It could be argued that Chatham was on a lifeline in the 1960s and 1970s; the size of the fleet was continuing to decline and political decisions would soon be taken to reduce the number of dockyards, Portsmouth and Devonport benefitted from better access to the sea for warships than Chatham which is located on the Medway River. The dockyard at Devonport had begun servicing nuclear-powered submarines about 1975 and the Submarine Refit Complex, built in the late 1970s, was eventually opened by the Prince of Wales in 1980.[6] Chatham eventually closed in 1984 and nuclear submarine refitting was transferred to Devonport Dockyard and it is now the only dockyard in the UK licenced to conduct nuclear submarine refuelling.

The Admiralty approached the United States for information on refuelling, with the Technical Chief Executive, Rowland Baker, visiting the US in May 1959. After discussions, enquiries were made by A. F. Honor, Staff Constructor Officer at the BJSM, into the facilities provided by the US Navy for the maintenance and construction of nuclear submarines. In a letter, SEO(P)/A.95/AO.1, dated 29 July, Honor referred to a visit to

[5] TNA, ADM 1/26843, HM Submarine *Dreadnought*: staff requirements, 1957–58.

[6] Peter Brimacombe, *Drake's Drum: A history of the Devonport Naval Base and Dockyard* (Plymouth, Mor Marketing, 1998), p. 56.

the Portsmouth Naval Shipyard, New Hampshire, which was engaged both in construction and refitting of nuclear submarines.[7] Honor advised that during the period the submarine pressure hull is opened to access the reactor compartment for refuelling and portable houses are secured on the pressure hull to provide clean working conditions and changing facilities for personnel. The core house had sliding roof panels to permit a crane lift through the house for lifting the core with its shielding.[8] Although the USS *Nautilus* had been in service for four years and other nuclear-powered submarines had been commissioned, many of the facilities at Portsmouth were only then being installed. It is apparent that the Admiralty was keen to learn from US Navy experiences early on in their programme so that they would be ready in all respects once the first nuclear-powered submarine had commissioned. This was especially so considering the greater potential hazards arising from refuelling (irradiated fuel rods) as compared with initial construction. The same year a film recording was made of *Nautilus*' refuelling and a copy was sent to the Admiralty for training purposes.[9] However, whilst the film would show the general handling requirements and procedures to refuel a submarine core, in this case the S2W core, it would not show the procedures required for the S5W Core 2 that was fitted in HMS *Dreadnought*.[10] Furthermore, in a memorandum reviewing access to submarine nuclear propulsion plant information, it noted that the lack of information was most keenly felt in the area of the reactor compartment especially in continued refit experiences and refuelling experience. The memorandum also noted that an additional complication was the belief that in her 1965 refit the USS *Skipjack* would be fitted with Core 3 presenting further limited access. The author wrote that as an example, any attendance or participation at *Skipjack*'s refuelling would: '… have to be confined to de-fuelling only and not to the insertion of the new core'.[11]

As part of their advance planning for *Dreadnought*'s first refit, Captain J. B. Morrison of the Dockyards Department, accompanied by Commander W. F. Moore, on the staff of the Commander British Navy Staff (CBNS) Washington, visited the shipyard at Charleston, South Carolina, in

[7] DNP 2, NP184/2011, p.83.
[8] DNP 2, NP184/2011, p.83.
[9] TNA, AB 6/1818, Note, Schonland to Wilson, 4 August 1959.
[10] TNA, AB 6/1818, Paper, H. Bowker, 3 February 1959.
[11] TNA, DEFE 72/45, Memorandum, B.N.S.00980/64, 9 September 1964, para. 10.

December 1962 to discuss arrangements being made to receive its first nuclear submarine refit, the USS *Scorpion*. The refit was completed in early 1964 and a request was ready to be made to the US Navy Bureau of Ships seeking a follow up visit by Commander Moore to access the shipyard's management experience in order for the Ministry of Defence (MoD (N)) to make their own decisions on personnel and material requirements.[12] However, in a letter to the Controller, CBNS advised that although the Chief of BUSHIPS, Admiral Brockett USN, thought the request was reasonable, Rickover was continuing with his strict interpretation of the extent to which the MoD (N) should have access to S5W plant operating information and could not be moved from opposing the visit.[13] However, Rickover was not as intransigent as many letters portray him to be. Rickover had called Sir Solly Zuckerman to remind him that *Dreadnought*'s refit was due to start soon and suggested drafting a letter to him anticipating the UK's need to draw on US experience.[14] Indeed, in the same letter Rickover advised Zuckerman that a refuelling manual was being prepared and would be made available to the MoD (N). Rickover also suggested that it would be most desirable that Rolls-Royce dealt directly with the US firm best placed to help, Electric Boat Division at Groton. Rickover sought the approval of the US Atomic Energy Commission to allow the termination of the Rolls-Royce/Westinghouse contract which would allow Rolls-Royce to negotiate a new contract with Electric Boat Division to provide the necessary information and equipment to refuel HMS *Dreadnought*.[15] It is important to note that Rickover used his position in the USAEC to drive forward his viewpoints on reactor design and as such was disliked by many senior officers in the US Navy who may have used Rickover as a convenient scapegoat when dealing with requests for nuclear propulsion information from the Royal Navy.

In April 1960, Rolls-Royce and Associates had written a technical memorandum on fuel handling inside the Dounreay reactor compartment. The Health and Safety branch of the UKAEA commented on the firm's proposals for unit module fuel handling noting that: 'The present

[12] On 1 April 1964 the Admiralty was dissolved and the Ministry of Defence (Navy) assumed its responsibilities. References from that date will be to MoD (N).

[13] TNA, DEFE 72/45, Letter, Vice Admiral Bush to Vice Admiral Le Fanu, 9 September 1964.

[14] TNA, DEFE 19/243, Letter, Zuckerman to Vice Admiral Le Fanu, 16 March 1965.

[15] TNA, DEFE 19/243, Letter, NR:D:HGR, Rickover to Zuckerman, 17 June 1965.

experience with this method of defuelling is very limited…'.[16] The memorandum further noted that a decision could not be made on the method to be adopted, the other method of fuel handling being core cartridge, until sufficient US experience was available for them to evaluate the two methods. It would appear from this memorandum that unit module replacement of cores was the preferred method for refuelling; however, by the time Rolls-Royce and Associates began negotiating with Electric Boat Division it was the UK's intention to use the core cartridge method. Zuckerman was advised that although the core cartridge method was technically feasible, the US does not use it and had not: '… developed the necessary procedures or equipment', and the US would not be able to provide assistance.[17] The note further advised that Electric Boat Division had already begun preparing the procedures and equipment for the module technique method employed in refuelling the USS *Skipjack*. The fact that the MoD (N)'s preferred method of refuelling differed from the US obviously required a re-evaluation of the methods, and details of the module replacement method were requested from Westinghouse. After studying the information supplied by Westinghouse, Zuckerman wrote to Rickover to confirm that the UK authorities would now use the module replacement method for refuelling *Dreadnought*.[18] A letter from the USAEC to Electric Boat Division authorising them to negotiate a contract with Rolls-Royce and Associates noted among the clauses that the contract shall terminate ninety days following the first hot criticality of *Dreadnought* following the first refuelling.[19] The Controller, Vice Admiral Horace Law, was disappointed with the terms of the contract being limited to *Dreadnought*'s refuelling but was aware that Rickover had given assurances to Zuckerman on information being passed affecting the operation and safety of the submarine.[20] Rickover had previously given his assurance to Zuckerman in a letter he received 17 June 1965, that any safety information found to affect *Dreadnought* would be provided to the UK.

With the contract in place, Rolls-Royce and Associates contacted the nuclear power department at Rosyth dockyard with an agenda for a meeting at the end of December to discuss a forthcoming planning meeting

[16] TNA, AB 8/1126, Technical assessment: Dounreay Submarine Prototype Reactor, 1960–61, Memorandum, Nks/Uni.1/MW.B26.13.2.61, 13 February 1961.

[17] TNA, DEFE 19/243, Note, CSA/193/04, 25 June 1965.

[18] TNA, DEFE 19/243, Letter, SZ/694/65, Zuckerman to Rickover, 20 October 1965.

[19] TNA, DEFE 19/243, Letter, Block to Jones, 12 October 1965.

[20] TNA, DEFE 19/243, Letter, Controller to Zuckerman, 27 October 1965.

with Electric Boat Division. Topics for discussion included the organisation at Electric Boat, the programme and timescales associated with refuelling, facilities and services, documentation, equipment, personnel, and training.[21] There was a lot of information to be requested from the US and little time remaining to implement any processes/equipment at Rosyth that may have been overlooked. It was envisaged that the refuelling equipment would be at Rosyth in October 1966 ready for *Dreadnought*'s planned refit to start in December and core change in March 1967.[22] However, during a routine survey of her hull, cracks were found in the internal bulkheads of HMS *Dreadnought* which, it was reported, would bring her refit forward by a year.[23]

The Strength of Steel

It was not only the construction of the reactor and the primary circuit that occupied the minds of the metallurgists. No longer limited by the requirement to surface or snort at periscope depth to run diesels in order to charge their battery, the advent of nuclear propulsion gave the true submarine freedom to roam the ocean depths in order to hide from adversaries. This enhanced freedom—to use the full three-dimensional aspect of submarine warfare—would require new steels with higher tensile strengths to withstand the greater pressures that a submarine would have to operate at. Prior to World War II, Royal Navy submarines were typically constructed from High Tensile Steel (HTS), which gave a yield strength of seventeen tons per square inch and the hulls were of rivetted construction; the early rivetted 'U' class had a diving depth of two hundred feet. It was only in 1940 that a high-strength steel that could be welded and was suitable for submarines became available. The 'S' quality steel had similar mechanical properties of HTS with elements of carbon, manganese and silicon but had a greater yield strength of eighteen and a half tons which would give a deeper diving depth for the same weight and dimension of a pressure hull, the welded 'V' class diving depth was stated to be three hundred feet.[24] However, it was not easy to weld, and care had to be

[21] TNA, ADM 328/29, Rolls Royce/Westinghouse contract, RRA/Electric Boat: Refuelling HMS/M Dreadnought, 1965–75, Letter, Pepper to McHugh, 16 December 1965.

[22] TNA, ADM 328/29, Letter, Ppr 1/DM, Pepper to Hedger, 29 November 1965.

[23] 'Earlier Refit for Dreadnought', *The Times*, 23 September 1965, p. 5, Issue 56434.

[24] D. K. Brown, 'Submarine Pressure Hull Design and Diving Depths Between the Wars', *Journal of Naval Engineering*, Vol. 30, Book 3, (Dec., 1987), 511–519 (p. 513).

taken. Construction of the rivetted 'T' class submarines began in 1938 and they had an operational depth of three hundred feet; their welded sisters, with a slightly thicker pressure hull, had an operational depth of three hundred and fifty feet. However, the newly designed 'A' class had an even greater operational depth of five hundred feet, its pressure hull being a tenth of an inch thicker than the welded 'T' class.[25]

Development of new steels continued and a manganese/molybdenum-tempered steel, known as UXW, was used in the post-war construction of the *Porpoise* class during the 1950s, further increasing operational depth; however, due to weldability problems and cracks occurring, a better steel was sought. In fact, so bad was the quality of the steel that the Controller noted in a memorandum: 'Nothing in this world would induce me to approve any further submarines being built in the new UXW steel'.[26] As a result, a new quenched and tempered steel, QT28 (twenty-eight refers to its tensile strength in tons), was developed and used in the construction of the follow-on *Oberon* class submarines and gave an improved diving depth of six hundred and fifty feet. The development of new steels also required investigation of new welding materials and techniques to fabricate the pressure hulls. Alongside the improvements in steels and welding were developments in non-destructive testing (NDT) methods to ensure welds were clean and free from defects, these tests are primarily conducted through ultrasonic and, where possible, radiographic techniques. In his article, Sims, who was Director General Ships at that time, noted that without successful development of welding techniques: 'the building of *Dreadnought* could not have been contemplated'.[27] Whilst QT28 steel further increased operational depth, it would not be enough for HMS *Dreadnought*, and an advanced HTS steel, QT35, with an increased percentage of some elements and the introduction of vanadium was developed for her construction.[28] However, this new steel would bring its own problems for *Dreadnought*.

After sea acceptance trials and commissioning in early 1963, *Dreadnought* undertook many trials to prove her various systems and was docked on regular occasions for routine surveys of the hull and external

[25] Brown, 'Submarine Pressure Hull Design…', p. 515.

[26] TNA, ADM 205/106, (8d) Memorandum, Edwards to Mountbatten, 27 September 1955.

[27] Alfred J. Sims, 'Some Post-War Developments in Naval Construction', *Journal of Naval Engineering*, Vol. 13, Book 2, (Dec., 1961), 174–205 (p. 185).

[28] Sims, 'Some Post-War Developments…', p. 183.

fittings. In the summer of 1965, Dreadnought underwent her third routine hull survey at Rosyth dockyard. Ultrasonic testing exposed hairline cracking behind a number of welds connecting the pressure hull with internal frames and bulkheads. Similar testing had been carried out in December 1964 during her previous hull survey and no cracks had been found.[29] There were a possible number of factors relating to the causes of the cracks: stressing arising from welding contractions, the cyclic changes in depth on the pressure hull leading to fatigue, and defective welding. It was noted during *Dreadnought*'s construction that the shipbuilder had experienced difficulty in making satisfactory welds and had used a number of different electrodes in trying to overcome the problem.[30] DG Ships considered fatigue to be the main cause of the cracks and a number of these were to be repaired whilst the submarine was docked. *Dreadnought* was scheduled to undertake more trials during the early part of 1966 and DG Ships proposed not to advance *Dreadnought*'s planned docking in December 1966. As a precaution, ultrasonic surveys were conducted on *Valiant* and *Warspite* to ascertain their condition which was not envisaged to delay their construction. It was known that the US had experienced similar cracks in the hulls of their submarines but it was not known if they had conducted repairs. A British team that had visited the US in July 1965 formed the impression that the US were not unduly worried by the cracks.[31] However, despite the pressure hull being designed with specific attention to strength against fatigue failure, the special welding techniques that had been developed and the extensive non-destructive testing conducted, the fact that the QT35 steel was causing much remedial work to be organised and performed was a concern to the MoD (N) that had to be addressed.[32]

There can be little doubt that much money was spent on rectifying *Dreadnought*'s many welding defects, but by mid-1966 moves were underway to purchase an American steel, HY80. It was reported that HMS *Valiant*, which at the time the letter was written had not been

[29] TNA, T 225/2663, Design, construction and modernisation of the Nuclear Submarine HMS Dreadnought, 1960–66, Letter, Bryars to Nicholls, 15 October 1965.

[30] TNA, T 225/2663, Letter, Bryars to Nicholls, 15 October 1965.

[31] TNA, T 225/2663, Letter, Bryars to Nicholls, 15 October 1965.

[32] Rowland Baker and L. J. Rydill, 'The Building of the two Dreadnoughts. European Shipbuilding: One Hundred Years of Change', *Proceedings of the Third Shipbuilding History Conference at the National Maritime Museum, Greenwich*, 13–15 April 1983 (National Maritime Museum, 1983), p. 19

commissioned: '...is also experiencing cracking problems'.[33] The letter to the Treasury concedes that the first three nuclear-powered submarines, and to a lesser extent the Polaris submarines, will face continuing expenditure on rectifying the QT35 cracking problems. However, there was an operational and financial advantage to be had in employing HY80 steel in hull construction on boats four and five, and possibly six and seven.[34] The MoD (N) argued that the two firms making QT35 steel, Colvilles and Consetts, accepted that the quality of their steel was not high enough and they were examining ways the steel could be improved. Pritchard was seeking Treasury authority to spend £280,000 ($800,000) on HY80 steel and this was at a time when British governments tried to keep expenditure in dollars to an absolute minimum.[35] The steel costs of QT35 for the first three submarine hulls was approximately £450,000 each (£600,000 each for the Polaris-equipped submarines).[36] There was an argument for postponing the construction of the submarines until British steel makers could supply HY80 steel; however, it was also noted that there would be research and development costs, subsidised by the government. If there was no commercial gain to be had by the UK steel industry and the only requirement for the new steel was for submarine construction: '...then the Navy should go on buying this special steel from America'.[37]

The MoD (N) was already putting the Treasury under pressure by seeking authority to purchase the long lead items for the sixth nuclear-powered submarine, which would include HY80 steel for the hull. Indeed, in the same letter, it would appear that the Royal Navy already had a programme to develop an even higher strength steel, QT45; however, this was possibly abandoned due to the problems of QT35 steel, the economic reality of MoD (N) having to fund development of a steel with limited applications and of American willingness to sell HY80 steel to Britain.[38] Economic discussions continued between the Treasury and MoD (N) officials and eventually authority was given to proceed with procuring HY80 steel for SSN06 (and SSN07 which was to be the first of the improved SSNOX,

[33] TNA, T 225/2964, Nuclear submarine construction programme, 1965–66, Letter, N/MAT.I/S.10280/66, Pritchard to Nicholls, 24 June 1966.

[34] Note: HY80 refers to High Yield steel with a tensile strength of 80,000 p.s.i.

[35] TNA, T 225/2964, Letter, Pritchard to Nicholls, 24 June 1966.

[36] TNA, T 225/2964, Letter, Bennett to Nicholls, 4 July 1966.

[37] TNA, T 225/2964, Letter, Bennett to Nicholls, 4 July 1966.

[38] TNA, T 225/2964, Note, 2DM.128/209/03, Bennett to Nicholls, 12 July 1966.

Swiftsure class).[39] The following year, Treasury approval was sought for the long lead items for SSN08, *Sovereign*, which included HY80 steel plating.[40] However, during construction of the *Swiftsure* class a new British steel was developed, NQ1, which was significantly stronger and cleaner than QT35.[41] During research, it has not been determined which was the first submarine to be constructed with this new material. This is possibly due to the fact that the relevant papers have not been released, as NQ1 steel is still used in constructing today's Royal Navy submarine pressure hulls.[42]

Core Development

With the knowledge that the US exchange of nuclear propulsion information would cease soon after HMS *Dreadnought*'s reactor had gone critical, the Admiralty immediately started looking towards future core development. The Admiralty had requested UKAEA assistance in operating the Dounreay prototype, examination of the fuel elements, consideration of engineering designs submitted to the Admiralty and general research and development support. The UKAEA Industrial Group, based at Risley, wrote Memorandum 687(R), which outlined their future work for the submarine reactor development programme. Stage I involved the start-up of HMS *Dreadnought*'s and Dounreay's reactors including safety assessments. Stage II of the programme was to develop a new core for Dounreay with increased power and longer life and to investigate alternative control materials to hafnium. Judging this last line of investigation, it can be assumed that the Admiralty still had concerns about future availability of stocks of hafnium, it is doubtful they were looking to reduce costs knowing that hafnium was a far more effective absorber than the other materials so far tested. It was proposed that Stage II work would include safety assessments, shielding work and the rebuilding of Neptune for in-pile experimentation; the estimated cost for this work was put at £1,198,000 between 1961 and 1964.[43] This equated to 114 years of manpower effort

[39] TNA, T 225/2964, Letter, Nicholls to Pritchard, 10 August 1966.

[40] TNA, T 225/3324, Nuclear submarine construction programme, 1966–69, Letter, N/S 1004/67, Swain to Patterson, 24 February 1967.

[41] P. G. Wrobel, 'U.K. Nuclear Submarines: The development of the overall design', *Journal of Naval Engineering*, Vol. 28, Book 1, (Dec., 1983), 106–119 (p. 114).

[42] https://assets.publishing.service.gov.uk, p. 4 [Accessed 4 December 2020]

[43] TNA, AB 6/2308, Memorandum 687(R) in covering Letter, Fletcher, 27 July 1959.

from the Industrial Group alone, which indicates a sizeable commitment from the UKAEA. Commenting on Stage II, Rear Admiral R. S. Hawkins, who was responsible for nuclear propulsion, considered that given the machinery at Dounreay is designed to take 20,000SHP, advantage should be taken of increasing core power as development proceeds. Hawkins also noted that insofar as the UK was aware, there was no current development in the US beyond the 15,000SHP of the S5W plant for submarines; therefore, assuming there is still a staff requirement to develop submarine plants of higher powers: '...there appears to be full justification for undertaking this work in the U.K.'[44] It was inconceivable that having produced so many types of reactor in a short period that Rickover would stop trying to improve the core or, indeed, build new types of reactors for other classes of ships. Rickover had in fact been given formal approval for a feasibility study into a reactor plant for an aircraft carrier in November 1951, which led to the keel laying of the world's first nuclear-powered aircraft carrier in 1958, the USS *Enterprise* commissioned in 1961 and fitted with eight A2W pressurised water reactors.[45] The A2W plant was a big ship reactor and more powerful than the S5W plant (35,000SHP), the eight A2W reactors fitted in the USS *Enterprise* gave a total output of 280,000SHP.[46]

As part of an initial assessment of future core requirements, staff of Director General Ships reviewed the case for the pressurised water reactor and concluded that the future programme should concentrate on this type of reactor. With the possible exception of the boiling water reactor, the pressurised water reactor was considered the only type presently available which was inherently simple and easy to operate.[47] In this paper, staff divided the proposed Stage II development into two parts, Stage IIA was to develop a core to enable the full power of the Dounreay machinery to be utilised. Stage IIB was more ambitious and aimed at investigating a longer life core and reduced water pressure possibly leading to partial boiling: '...and developing towards the boiling water reactor'.[48] Both the pressurised water reactor and the boiling water reactor are light water

[44] DNP 6, CP60801, Memorandum, M.I./603/22/59, 30 July 1959.

[45] Hewlett and Duncan, *Nuclear Navy*, p. 401.

[46] https://www.globalsecurity.org/military/systems/ships/systems/a2w.htm [Accessed 18 September 2020].

[47] TNA, AB 6/1818, Memorandum, Submarine Nuclear Machinery Development, April 1960.

[48] TNA, AB 6/1818, Memorandum, Submarine Nuclear Machinery Development, April 1960.

reactors with advantages and disadvantages associated with each; however, in principle they operate in much the same manner. The pressurised water reactor uses the primary coolant to heat the water in the steam generator from where the steam is piped, through the secondary system, to the turbo-generators and main turbines before being condensed and fed back to the steam generator. The primary system and secondary system associated with the pressurised water reactor plant are independent of each other, whereas the boiling water reactor does not have separate systems. The boiling water reactor is designed to allow the water to boil in the core, the resulting steam is directed to the turbine to generate electricity before being condensed and pumped back to the reactor.[49]

One may argue that the pressurised water reactor is the better proposition in a submarine because the primary circuit is enclosed within the reactor compartment. In contrast, the boiling water reactor would produce steam in hitherto secondary systems in the machinery spaces, thus making containment of any accident that more difficult to achieve. It would be prohibitive to site all the secondary parts of the steam machinery (the turbo-generators, main turbines, condensers and feed pumps) within an extended reactor compartment boundary; it would mean more automation just as the Admiralty was trying to reduce the number of remotely controlled valves in the reactor compartment. Baker had highlighted the boiling water reactor's unsuitability to submarine steam plants because of the contamination of the air ejector system, he also noted that the designer would have to overcome the inherent instability of a rolling and pitching submarine which, in extreme circumstances, may uncover the fuel elements thus leading to possible overheating of the fuel elements.[50]

In a memorandum to the Controller, the Director General Ships, Sir Alfred Sims RCNC, envisaged that a future programme would develop a competent body of engineers and scientists and prove the UK's ability to provide a longer life core with greater output. The development programme would be conducted in cooperation with the UKAEA and Rolls-Royce and Associates, which would utilise knowledge gained from Westinghouse, Sims stated that if the core life was doubled it would save £6 M throughout the life of each submarine.[51] In effect it would mean longer periods between refitting a submarine and a reduction in the

[49] Davies, *Glossary of Nuclear Terms*.
[50] TNA, AB 6/2308, Letter, RB/144/60, 13 April 1960.
[51] TNA, ADM 1/27750, Memorandum, S.10007/60, 27 September 1960.

refuelling costs associated with docking the submarine. If the Treasury sanctioned the programme it would necessitate rebuilding the zero energy reactor Neptune, which was defueled and dismantled when the decision was taken to purchase the S5W plant. Neptune was stored by Rolls-Royce at their Derby premises, the estimated cost to rebuild, fabricate a new core and procure fissile material totalled £3.9 M during 1960/1961. The Controller, Admiral Sir Peter Reid, expanded on and formally submitted the memorandum to the Admiralty Board asking a series of rhetorical questions considering the aims of a five-year Core Development Programme. Whilst focussing on the requirement for the UK to develop its own core and establish a body of nuclear knowledge, the Controller envisaged that at the end of the programme the UK would have closed the gap in its competency.

The conventional propulsion machinery for Dounreay and HMS *Valiant* (and follow-on submarines) was designed to develop 20,000SHP in contrast to the conventional machinery of the *Skipjack* class, which was geared to the S5W plant power output of 15,000SHP. By developing a core to match the potential of the conventional machinery, the Controller reasoned that the exercise would: '...give us the assurance of our ability not only to design a P.W.R. plant but to design a reactor to meet a specific power requirement'.[52] The Admiralty Board gave their formal approval in mid-October but urged that pressure should be brought to bear on Rolls-Royce to contribute financially, either directly or indirectly towards the development costs. It was also suggested that the UKAEA be investigated for possible contributions to support development as they too would benefit from the programme.[53] By successfully developing their own core, the Admiralty would not only justify Rickover's judgement in the UK's ability to stand on its own merits in nuclear engineering, it would also release the Royal Navy from the power constraints of the S5W reactor, albeit the operational advantages of the increase in power would be minimal. Most important, one may argue from their viewpoint, would be the possibility for Admiralty scientists to return to the US within a few years with their knowledge and experience and attempt to reopen the doors of nuclear reactor information exchange.

[52] TNA, ADM 1/27750, Memorandum, B.1362, October 1960.
[53] TNA, ADM 1/27750, Board Minute 5447, 20 October 1960.

Amended Access Agreements

With the signing of the Westinghouse/Rolls-Royce contract, 10 February 1959, the UKAEA's participation in the nuclear submarine programme had altered. To clarify their position and their future participation, Baker convened a meeting to discuss the UKAEA's role. Baker thought the Admiralty and the contractors could explain to the UKAEA their methods of working and the division of responsibility in the nuclear submarine programme. All the parties could then discuss the part the UKAEA could play and agree upon their role. The UKAEA considered their role was to give a service to the Admiralty and not give direction to Admiralty contractors.[54] The UKAEA would examine the engineering proposals for the Dounreay plant, comment on them and make a safety assessment of the plant, they would also assist the Admiralty in setting up the operations team and would provide training within the scope of their facilities. The UKAEA would also provide for examination of fuel elements within the existing facilities at Dounreay and would advise on facilities required for reprocessing used cores and reprocess them if required. Baker advised that for the time being, the US had considered it not worthwhile reprocessing used cores but thought they might start within the near future. It was agreed at the meeting that the UKAEA would investigate the disposal of irradiated cores should reprocessing be considered not worthwhile. This was a financial consideration, there had been discussions on the practicality of reprocessing used cores since 1957, and it was known that the US had successfully completed reprocessing a zirconium core that year and that the procedures for doing so would be made available to the UK.[55]

In the earlier stages of research and development, Access Agreements had been signed in 1956 between the UKAEA and the firms involved; however, since the Westinghouse Agreement, Rolls-Royce considered that there was no necessity for them to enter into an Access Agreement with the UKAEA for the proposed future development. In 1956, the UKAEA was the sole (British) source of nuclear experience and information, Rolls-Royce and Associates now argued that they too had nuclear experience stemming from their Westinghouse contract.[56] In reply, Baker's deputy, Captain Terence Ridley, thought Rolls-Royce was taking a narrow view

[54] DNP 6, CP60801, Minutes, 4 December 1959.
[55] TNA, AB 16/1380, Letter, Lang to Strath, 24 October 1957.
[56] DNP 6, CP60801, Letter, Hedger to Baker, 19 November 1959.

trying to protect Westinghouse information from the UKAEA which, Ridley argued, could be written into any future contract or Access Agreement. '...whatever one may think of A.E.A., their peculiar position makes it inconceivable that Rolls-Royce and ourselves could ignore them entirely for the next ten years'.[57] Rolls-Royce expressed the view that the caveats concerning design rights and patents included in the original 1956 contract, C.P.101/55, should not apply to the Dounreay core and should cease after signing the Westinghouse Contract in February 1959. The Admiralty would not accept these conditions and insisted on their retention in both contracts.[58] Rolls-Royce's senior chief executive co-ordinating the nuclear effort, H. L. Barman, thought that these should be renegotiated. In connection with the Access Agreements, the Admiralty wrote to the three firms involved to advise that the UKAEA's participation in the nuclear submarine development programme needed to be revised following the assistance given by the US, which necessitated new Access Agreements being concluded. The letter gave details of the UKAEA's future participation and a specimen of the Access Agreement was attached for the firms' approval.[59]

The first part of the letter was sent to Barman at Rolls-Royce, R. P. H. Yapp at Vickers-Armstrongs and C. E. H. Verity at Foster Wheeler. The second part was addressed solely to Rolls-Royce and referred to the firm's concerns highlighted in Barman's earlier letter; the Admiralty reiterated that information passed to Rolls-Royce and Associates by Westinghouse was a result of that firm acting as the UK Government's agents and the disposition of design rights between the Admiralty and the UKAEA was a matter for them. Finally, Rolls-Royce's reference to the possibility of entering into a commercial licence agreement with Westinghouse was unlikely, as the last time it was proposed, the US Government was not in favour of it. There were to be many letters and drafts of the Access Agreement before the wording was agreed on and acceptable to the UKAEA and Rolls-Royce. It would appear from the letters that both the UKAEA and Rolls-Royce were intent on getting the other party to acquiesce to their version of the patents and design rights. Captain Ridley noted sardonically: 'Many thanks for the latest instalment in this thrilling serial [...] when I try to read the latest draft the words

[57] DNP 6, CP60801, Letter, Ridley to Hedger, 2 December 1959.
[58] DNP 6, CP60801, Letter, Hedger to Barman, 5 April 1960.
[59] DNP 6, CP60801, Letter, Hedger to Barman, Yapp and Verity, 25 October 1960.

swim before my eyes and become meaningless'.[60] It is clear that towards the end of negotiations people were becoming weary with the process of rereading draft copies. Eventually, however, the final draft was sent to Rolls-Royce and the UKAEA for signature at the end of December 1961.

Core Development Programme (CORDEP)

The main reason behind the research and development programme known as CORDEP was the Admiralty's intention to make good the gap in their knowledge of reactor design caused by the purchase of the S5W plant. In a letter to the Treasury it was advised that the result of this "corner cutting" would leave the Admiralty dependent upon the US—Admiral Burke USN, Chief of Naval Operations, and Rickover could change their minds at any time, and they would not be in office forever. It was also possible that the US Government could make things more difficult as administrations change.[61] Despite a limitation of ten years being written into the 1958 Mutual Defence Agreement, in this and other papers researched the Admiralty appear to stick to the possibility that there would be an opportunity to continue or renegotiate the Agreement. However, it was always Admiralty policy to proceed independently once all possible information had been gained from the 1958 Agreement. The Admiralty's continuous references to "possible further assistance from the US" must, therefore, be seen as a means of demonstrating to the Treasury that efforts were still being made to save expenditure on research and development by gaining future assistance from the US. Knowing that future research would be conducted independently was the rationale behind the CORDEP programme as it was only through independent endeavour that the Royal Navy would manage to build its own nuclear submarine fleet. After deliberation by the various Admiralty departments, Admiralty Memorandum S.10007/60 was ready for submitting to the Treasury in early 1961. However, the Under Secretary of Finance at the Admiralty, J. M. Mackay, was not content with the memorandum and sought to change the emphasis prior to presenting it for approval. Mackay thought the memorandum protested too much about the Admiralty's right to a research and development programme and dwelt too much on the thorny issue of "dependence on America". Mackay argued that the memorandum: '…instead of allaying

[60] DNP 6, CP60801, Letter, D/TCE.729/61, 14 December 1961.

[61] TNA, ADM 1/27750, Memorandum, Core Development Programme, November 1960?

criticism, may well stimulate attack, especially with the incentive to save £10M'.[62] Mackay proposed to submit a memorandum which focussed on the resumption of the research and development programme that was curtailed which would complete the Admiralty's pressurised water reactor knowledge, produce worthwhile economies and put the Admiralty in a position to trade with the Americans. Lang agreed with Mackay that his alternative draft should be put to the Board for their consideration.

Mackay's version of the memorandum was eventually forwarded to the DRPC for their consideration 26 April 1961. It appears that the DRPC advised the Admiralty to contact the US for advice on the proposed programme. The Minister of Defence, Harold Watkinson, wrote to the US Secretary of Defense, Robert McNamara; similarly, the First Sea Lord, Admiral Sir Caspar John wrote to Admiral Arleigh Burke USN. The Admiralty received interim replies to the letters which agreed that the programme was a sensible proposition; indeed, the letter advises the Treasury that: 'The U.S. Navy in particular have given assurance of their support, and there is no reason to think that the final U.S. reaction will be any different'.[63] It is possible that these letters were sent in order to satisfy the Treasury that efforts were still being made to elicit assistance from the US but, inevitably, the man who would ultimately advise McNamara would be Rickover, and his views on future assistance were well known to the Admiralty. The US Navy would be able to offer very little support if the US Government did not also sanction some form of assistance programme. Acknowledging that they were waiting on a final reply from McNamara, the Admiralty sought approval for six months of funding estimated at £3 M, which included costs for fissile material.

By reply, the Treasury thought there were arguments for starting the programme but what worried them was the lack of indication of how much assistance the Americans were prepared to offer that may reduce costs and increase efficiency. Reluctantly, the Treasury agreed that the Admiralty could make a start on the project, limited to £2.15 M, on the understanding that their commitments were kept to a minimum until the US made known its policy on further assistance.[64] The Ministry of Defence responded to the Treasury's concerns by noting that they had drafted a

[62] TNA, ADM 1/27750, Minute, Mackay, 1 February 1961.

[63] TNA, ADM 1/27946, Machinery for nuclear submarine: future research and development, 1961–67, Letter, S.3630/61, 27 July 1961.

[64] TNA, ADM 1/27946, Letter, 2-DM.209/32/02, 14 August 1961.

letter for Sir Solly Zuckerman to send to the US enquiring when we might expect further information concerning the position of the USAEC and US Navy to our plans: '...alias in both cases Admiral Rickover'.[65] Despite a number of letters expediting a formal response from the US the reply from McNamara on 9 April 1962 was not promising. The USAEC Chairman had informed McNamara that based on the information provided, the proposed UK programme of research and development: '...should significantly assist in increasing your competence in the nuclear propulsion field'.[66] This is a very non-committal statement from the US and is a reiteration of the main intention of the 1958 Agreement, namely that the UK would stand independently and progress from the knowledge transferred under that agreement. Despite McNamara advising that the letter was an interim response, he thought that: '...something more definite will reach us before long'.[67] No further reference was found to McNamara's reply and it can be assumed that Rickover refused all further requests for assistance or advice, indeed, Baker observed that: '...we have gained enormously from the Agreement [...] but all we seem to get now or seem likely to get in the future are "restrictive fringe effects"'.[68] In fact, it is recorded in the same Minute that the Admiralty commenced the Polaris programme on the basis of their existing nuclear knowledge. 'During the negotiation of the Polaris Sales Agreement we were warned by the Americans that they could give us no further nuclear information'.[69]

The Treasury, the Cabinet and certain elements within the Admiralty began to face the reality of the 1958 Agreement and resign themselves to the fact that there would be no further assistance from the US. A joint communique was released by the US and UK Governments 21 December 1962 announcing the Nassau Agreement by which the UK was to purchase the Polaris missile system from the US; after negotiations the Polaris Sales Agreement was signed 6 April 1963. From the autumn of 1962, in the files researched there appear to be no further references from the Treasury to the Admiralty concerning US assistance in their nuclear propulsion programme. I would argue that from the signing of the above

[65] TNA, ADM 1/27946, Letter, S.3630/61, 16 August 1961.

[66] TNA, ADM 1/27946, Letter, 2-DM.209/32/02, 11 April 1962.

[67] TNA, ADM 1/27946, Letter, 2-DM.209/32/02, 11 April 1962.

[68] TNA, DEFE 72/45, Register No. S.10583/63, Minute Sheet 1, paragraph 5, 20 August 1963.

[69] TNA, DEFE 72/45, Register No. S.10583/63, Minute Sheet 2, paragraph 3f, 3 September 1963.

Agreements, it was political determination to proceed with the Polaris deterrent, which necessitated nuclear-propelled submarines as the firing platform, that the Admiralty was able to proceed with their CORDEP programme of research and development without further requests from the Treasury concerning future US cooperation.

Nuclear Development (Submarines): NuDe(S)

In December 1961, the finance branch at the Admiralty wrote to the Director General Ships, Sir Alfred Sims, requesting he put a figure on the extent of Admiralty commitments to the NuDe(S) programme at the end of March 1963.[70] This is the first reference found in the files researched to the NuDe(S) programme; confusingly, it is used in conjunction with CORDEP until the final reference to the latter programme in September 1963.[71] As discussed previously, Stage I of the development programme included the construction and operation of the UK-manufactured S5W core (Core A) in the Dounreay prototype; Stage II (now referred to as NuDe(S) I) related to the future development of an improved reactor core with higher power output and longer life. To achieve this objective, the Admiralty's proposal was: '...to increase the length of the core, increase the amount of fissile material and incorporate improved alloys and shielded burnable poisons'.[72] Burnable poisons are an effective neutron absorber and are used in nuclear engineering to reduce initial reactivity, they become less effective as core burn-up progresses, thus counteracting the fall in reactivity as the fuel is used up.[73] Begun in 1961, the programme required the Core B design to be agreed upon and frozen by mid-1965. This date would allow the programme to fit in with the predicted burn-up of Core A in the Dounreay prototype and would allow time to manufacture a core for Dounreay. The UKAEA was confident if no further delays were experienced in the experimental programme, then the data from the intermediate and final non-destructive inspections should provide confidence in the selection of the Core B fuel element parameters.[74]

[70] TNA, ADM 1/27946, Letter, Mawer to D.G. Ships, 18 December 1963.

[71] In correspondence, both programmes use the Admiralty reference S.3630/61 and the Treasury reference 2-DM.209/32/02.

[72] TNA, AB 16/3287, Nuclear ship propulsion: naval vessels including submarines, 1956–63, Note, A.E.A.(61)16, 24 February 1961.

[73] Walker, *Chambers Dictionary of Science and Technology*, p. 155.

[74] TNA, AB 6/2492, Memorandum, ASPWP/P21, paragraph 3.2.4, 26 February 1963.

Although the development programme had been given limited financial authorisation to proceed by the Treasury in October 1961, uncertainties concerning Admiralty estimates of about £10 M had prevented Treasury approval of the overall programme estimate. In the spring of 1963, the Admiralty wrote to the Treasury advising that the total estimate of the programme had fallen to £9.89 M and enclosed the expenditure to date and estimated costs going forward. This information satisfied Treasury concerns and approval was given.[75] However, as with most estimates the figures are liable to fluctuations; in their next report to the Treasury the Admiralty had to advise that due to wage increases and a rise in fuel costs the estimates had risen by £460,000, although they did point out that this figure was still within the original £10.5 M envisaged.[76] By August, however, the cost had risen to £10.74 M due to price changes in fissile material. In December 1966, it was reported that Core A had achieved a seventy percent burn-up, allowing time for refit and core change; it was envisaged: '…that Core B would go critical in the early autumn of 1968'.[77] Core B was installed at Dounreay during 1967 and initial criticality at zero power was achieved in June 1968. It was taken critical at low power in August and then operated at high power for the first two years to give confidence in the design prior to manufacturing the first production core for HMS *Swiftsure*, which had been ordered in October 1967.[78] Core B objectives of a greater power output and a longer life were achieved, the core operated for just under four years before final burn-up, a substantial increase on the two years achieved with Core A. It is worth noting that if the core was controlled under normal operating conditions in a submarine, rather than Dounreay's "proving" conditions, its operational life would be greatly increased above the four years achieved in the prototype.

The MoD (N) had decided early in the NuDe(S) I programme that so long as there was a nuclear submarine fleet there would be a requirement for a supporting research and development programme. In early 1965, the MoD (N) wrote to the Treasury with a progress report of the NuDe(S) I programme and to advise that they had decided to provide funds from this programme for a team of three Rolls-Royce and Associates personnel to:

[75] TNA, ADM 1/27946, Letter, 2-DM.209/32/02, 3 May 1963.
[76] TNA, ADM 1/27946, Letter, S.3630/61, 9 March 1964.
[77] TNA, ADM 1/27946, Letter, 14.B.1(A), 8 December 1966.
[78] Horlick, 'Submarine Propulsion in the Royal Navy', p. 12.

'...maintain a continuous review of the whole NuDeS programme'.[79] Funding of £54,000 was allocated in July 1964 to focus on noise reduction by examining two possible types of cores in the new reactor plant and to provide a planned and costed programme on which to base proposals for the development of such a plant. The MoD (N) sought Treasury sanction of this preliminary investigation to the end of December 1965 at a total cost of £145,000 covered by the existing £10.74 M budget. Much to their alarm, the Treasury replied: '...it seems to us that your programme is tantamount to a project study for a major new programme'.[80] The Treasury requested that the MoD (N) give an assurance that the investigations referred to in their letter implied no commitment to a new development programme. The MoD (N) was able to confirm that their continuing investigations did not commit them to a new programme; however, in a follow-up letter they did advise that nothing should be done to hinder the Navy in making the best judgement in reducing noise levels. The MoD (N) recognised that when proposals were submitted they would require careful scrutiny in consideration to the budget available and the possibility of modifying submarines already in service.[81] The Treasury, satisfied with the information provided, sanctioned their proposals, but by January it was envisaged that the investigations would not be completed until March with costs now expected to be £200,000.[82] The letter also gave the Treasury an outline on the status of the preliminary proposal for a new research and development programme which had been considered by the Board. The proposals had to be finalised and the investigations would need to be completed; the MoD (N) advised the costs could still be met from the allocated budget and sought the Treasury's approval to continue. Although the Treasury agreed to the proposed increase, there was to be another increase in the final estimate for the NuDe(S) I programme. Director General Ships sought £11.5 M to cover wage increases and the delays caused by the introduction of the Polaris programme; he also wanted to avoid any impediments to the contractor's work.[83] The Treasury, replied that if the MoD (N) could justify the importance of finding the extra expenditure within the existing defence budget, they would sanction the £11.5 M to complete NuDe(S) I.[84]

[79] TNA, ADM 1/27946, Letter, S.3630/61, 2 April 1965.
[80] TNA, ADM 1/27946, Letter, 2-DM 209/32/02, 9 April 1965.
[81] TNA, ADM 1/29746, Letter, RP/311/65, 18 May 1965.
[82] TNA, ADM 1/29746, Letter, S.3630/61, 14 January 1966.
[83] TNA, ADM 1/29746, Paper, NRDB/P(66)33, 18 October 1966.
[84] TNA, ADM 1/29746, Letter, Patterson to Heslop, 23 December 1966.

NuDe(S) II—FLIP—ANP

From the preliminary investigations being conducted under NuDe(S) I finance, it was apparent that the MoD (N) was considering two designs for their future programme to develop Core C. Option one, an integral pressurised water reactor, designated NR3, and option two, designated NR4, a supercritical water cooled system. Both options would have relied on natural circulation for cooling, although for NR3 it would only be adequate for power outputs up to thirty percent. Although the main objective was a reduction in noise levels, also to be included were economic incentives, development potential, ease of maintenance and simplicity in refuelling.[85] The MoD (N)'s proposals had been endorsed by the Nuclear Sub-Committee of the Weapons Development Committee, a draft submission from the Deputy Chief Scientific Advisor (Projects) to the Secretary of State for Defence, Denis Healey, seeking his agreement to commence Phase I was circulated for endorsement or comment by the Chiefs of Staff Committee Secretary.[86] The draft submission observed that the S5W plant was ten years old and it was now possible to consider radical improvements in a number of areas, such as performance and noise. It was estimated that Phase I of NuDe(S) II would take three years at a cost of £5 M which would cover three designs. The first two have been briefly described above, the third design originated from the AERE, which envisaged using plutonium cermet fuel plates, this option was being considered because it was thought that the UK's current stockpile of enriched uranium for use in reactor cores would be exhausted within ten years. The cost in reopening the high enrichment diffusion plant at Capenhurst and the possible problems in procuring fissile material from the US stemming from pressure to place the export of nuclear material under international safeguards meant the UK was looking at means for overcoming these possible difficulties. Another factor was possible costs; it was noted in the draft that plutonium fuel costs could be fifty percent lower than enriched uranium costs.

In *Rolls-Royce: the Nuclear Power Connection*, Lambert wrote that the proposal for Core C did not proceed beyond the preliminary design stage because attempting to achieve all the objectives set for the plant was

[85] TNA, AB 89/75, Dreadnought nuclear submarine project: organisation, 1965–68, Letter, O'Driscoll to Murray, 22 September 1965.

[86] TNA, DEFE 4/204, Chiefs of Staff Committee: Minutes of Meetings 40–46, 1966, Paper, COS 1826/8/8/66, 8 August 1966.

deemed too great and is dismissed as remaining as: '…no more than a paper proposal.'[87] One file researched indicates that Core C was more than merely a paper proposal, the work submissions to the MoD (N) in support of the NuDe(S) II contract amounted to nearly £1.25 M over three years between 1966 and 1969 covering all aspects of the steam raising plant from fuel technology to improving the properties and quality of the reactor pressure vessel. The work was conducted on behalf of the MoD (N) between Rolls-Royce and Associates and the Yarrow Admiralty Research Department (Y-ARD).[88]

A report by the Defence Equipment Policy Committee (DEPC) gives an account of the future development programme to October 1973 which ultimately sought funding for the design of a new plant, provision of a second prototype at Dounreay and continuation of the Secondary Plant Improvement Programme (SIP).[89] A reference in the report is made to a MoD (N) paper, WD(NS) 1/68, which recommended in addition to the programme of improvement of the present pressurised water reactor, known as the Primary Plant Improvement Programme (PIP), that effort should be devoted to defining the role and material parameters of a new generation of submarines (dubbed SSN0Y), the future *Trafalgar* class.[90] Another recommendation was to continue investigatory work for a new design of propulsion plant on the lines of NuDe(S) II. This programme was subsequently renamed the Forward Looking Investigatory Programme (FLIP) in 1969 and was renamed once more the Advanced Nuclear Plant (ANP) in April 1972. Another MoD (N) paper quoted in the DEPC report, DEP 28/72, sought endorsement of the doctrine that the saturated water reactor was the best form of future plant, that a new shore prototype was essential to the development programme and that funding of the design programme should be continued. Approval had been sought for an expansion of the FLIP programme for a period of three years, 1972/3–74/5 and for the SIP programme to be extended by one year to

[87] Lambert, *Rolls-Royce: the Nuclear Power Connection*, p. 56.

[88] TNA, AB 64/280, Rolls Royce and Associates (RRA) and Y-ARD work submissions: NUDE(S) II design and development programme for naval reactor plant, 1966–69.

[89] DNP 4, NP186/2011, Naval reactor programme and research and development. Report, N/S0011/72, The Naval Nuclear Propulsion Development Programme: Design of new propulsion plant, provision of second shore prototype, and continuation of Secondary Plant Improvement Programme, October 1973.

[90] SSN0Y project resulted in the *Trafalgar* class of submarine, three of which are still in service.

1974/5. The ANP programme entailed investigating a saturated water reactor plant capable of producing 40,000SHP; the associated SIP programme was preparing secondary machinery and equipment which would meet the new reactor requirement.

The DEPC report noted that further development of Core B was clearly a possible alternative to the development of the saturated water reactor but cautioned that further improvements to noise and shock vulnerability was limited. The cost of pressurised water reactor development was dependent on the maximum power requirement and the following systems were identified: a) a ten percent increase in power (20,000SHP) with no prototype testing, estimated development cost £18–£20 M; b) a ten to forty percent increase in power (25,000SHP) prototype testing using the existing DS/MP, estimated development cost £35–40 M; and c) a greater than forty percent power increase with testing conducted at a new shore prototype, estimated development cost £50–55 M.[91] Development of the saturated water reactor of whatever power and a new shore prototype was estimated to cost £60–65 M, which was clearly a more expensive programme for the MoD (N) than the PIP programme which was to become Core Z. The preferred option of the Future Generation Submarine (FGS) team investigating the requirements for SSN0Y was a power output of 25,000SHP; however, the report concluded that whilst the saturated water reactor was the more expensive option, it had considerable long-term advantage and potential for further development. It would give increased power, enhanced nuclear safety, reduced noise and reduced vulnerability to shock. The report envisaged freezing the main design parameters, health physics and reactor compartment layout in April 1974, core design freeze and order mid-1977 and core load mid-1980. There is a great amount of detail in the report with technical comparisons made between the two plants (Annex E), which determined that the greater advantages lay with the adoption of the saturated water reactor. The report concluded that the FLIP programme has achieved its objective in identifying the reactor system most suited for marine propulsion in the 1980s and invited the Naval Projects Board (NPB) to endorse the DEPC report that recommended future development be based on the saturated water reactor plant. As such, approval was sought to extend the ANP programme to July 1984 involving an expenditure of £55.2 M from

[91] DNP 4, NP186/2011, Paper, N/S0011/72, paragraph 21, October 1973.

April 1974. Approval to extend the SIP programme to April 1978 at a cost of £7 M from the same date was also sought.

In November 1973, Rolls-Royce and Associates placed a contract, NA/3/2107/900/Sqs, with the Yarrow Admiralty Research Department (Y-ARD) for the design of a secondary machinery plant suitable for an ANP of integral design.[92] The prototype plant would be known as DSMP 2 and sited adjacent to the present Dounreay prototype plant. The contract specifically called for a design which would be fully representative of a submarine secondary plant of similar power to allow predictions to be made of plant performance and chemistry effects. It was agreed that the proposed design should closely follow the basis of the Advanced Secondary Plant Reference Design (ASPRD) currently being progressed under the SIP programme. This work progressed until April 1976 when MoD accepted Rolls-Royce and Associates' advice to terminate work associated with the integral reactor plant and direct effort towards improvement of the dispersed pressurised water reactor (PWR 1) fitted in current submarines. The termination report provides details of the status of the system design in May 1976 but affords no reasons for the cancellation of the ANP saturated water reactor programme; another file associated with this work, ADM 317/205, is closed, and attempts to gain access were unsuccessful; the file is now listed as "officially missing".[93]

Core Z

Core Z, the final core update for PWR 1, was the result of the Primary Plant Improvement Programme (PIP) which had two objectives defined by MoD (N), to increase core life and to improve main coolant pump performance. In January 1969, Rolls-Royce and Associates published cost and timescale estimates, for the PIP contract intended to replace the NuDe(S) II contract.[94] The design of Core Z took three years, and nine variants were analysed before the mock-up was tested in Neptune; the design was agreed upon in mid-1972 to allow manufacture and delivery to Dounreay 8 April 1974. Core Z went critical for the first time 16 December 1974; however, during testing: '…some performance issues were

[92] TNA, ADM 317/204, Advanced Nuclear Plant (ANP) prototype secondary plant: termination report, 1976.

[93] Private correspondence, FOI 2017/03919, 26 April 2017.

[94] Lambert, *Rolls-Royce: the Nuclear Power Connection*, p. 63.

identified, which invalidated the full-power safety case'.[95] A major programme was undertaken to understand the problem and determine the means for resolving it, there were also implications for submarines in commission with Cores A and B. A solution was forthcoming and fitted to Core Z at Dounreay, and over the coming years the modification was fitted to the operational submarines. The burn-up of Core Z began in earnest in June 1976 and was finally shut down on 20 June 1984: '...having exceeded its expected lifetime by a useful margin' (Fig. 7.1).[96]

It must be appreciated that Core Z is fitted to three submarines still in commission, and as a result little or no information has been released concerning the PIP programme. Many of the NuDe(S) II files and those of the PIP Programme in the National Archives catalogue are advised as

Fig. 7.1 HMS/m *Triumph* engineering space (still cramped conditions)

[95] Lambert, *Rolls-Royce: the Nuclear Power Connection*, p. 64.
[96] Lambert, *Rolls-Royce: the Nuclear Power Connection*, p. 65.

closed and retained by the Nuclear Decommissioning Authority. From the manufacture of the first Dounreay core through to the success of Core Z, and beyond to PWR 2, a small cadre of reactor core designers have increased the power output and extended the life of a core constrained by the physical limitations of the reactor pressure vessel. As Hill noted, they have done a superb job which: 'because of security constraints, will only be fully appreciated by a small number of people in the company and in the MoD'.[97] HMS *Triumph*, the last of the 'T' class submarines, was commissioned in 1991 and was refuelled in 2005. She is due to decommission in 2022; that is surely enough evidence of these unknown engineers' skill in designing a reactor core that was safe to operate and "built to last".

Secondary Machinery Improvement

HMS *Dreadnought* had commissioned with a proven machinery plant, not only the S5W reactor, but also the propulsion plant and auxiliaries. Hammersley has noted that the *Skipjack* systems on HMS *Dreadnought* were a lot less complex than the machinery fitted to HMS *Valiant* and the follow-on SSN submarines and the SSBN *Resolution* class. Hammersley thinks that HMS *Valiant*'s crew and later teams: '…deserved great credit for coping successfully with the more complicated unproven British plant and without our experience'.[98] Britain had built steam-driven submarines in the past, designed with a surface speed of 24 knots and able to operate with the Grand Fleet; however, the 'K' class submarines were an unmitigated disaster. In service between 1916 and 1931, a number of them were lost to accidents, "The Battle of May" being the most infamous. Two submarines, K4 and K17, were sunk and four badly damaged as a result of collisions on the night of 31 January 1918. Being propelled by steam on the surface, they also had more hull openings than contemporary submarines, as one contemporary officer noted, there were: '…too many damned holes'.[99] There is hearsay evidence that Rickover acquired the machinery plans for the K class in order to get an understanding of how to fit a steam plant into the USS *Nautilus*, it is sufficient to note that on both sides of the Atlantic there was plenty of experience in steam-driven warships but the knowledge for steam-driven submarines was short-coming. It is,

[97] Hill, 'Forty years on:', p. 16.
[98] Private correspondence, Rear Admiral Peter Hammersley, 15 November 2018.
[99] Anthony Preston, *The World's Great Submarines* (Leicester, Blitz Editions, 1998), p.37.

therefore, not surprising to hear of the Dounreay secondary machinery plant being described as a shore-based *Leander* class frigate. These frigates were an improvement on the Type 12 frigate starting with the *Whitby* class and the modified *Rothesay* class; the Y.136 steam machinery plant in the *Leander* class was a well-designed and proven plant and was an obvious starting point for the nuclear submarine steam plant as built at Dounreay (Fig. 7.2).

Many of the auxiliaries fitted in the original five SSNs and the later *Resolution* class SSBNs were steam powered. Two of the four feed pumps fitted were turbo-driven and used only when the submarine went to the full power state for high speed; the disadvantage of using this system was the over-speed trips needed testing before taking the plant to the full power state, which took time, so full power was not immediately available in the event of an emergency. The turbo-driven feed pumps also shared the same supply and exhaust as the turbo-generators and there was a risk

Fig. 7.2 Swiftsure class submarine HMS/m *Superb* at sea 07/12/1976

of collapsing the turbo-generator vacuum when it was needed most. All the machinery in these classes of submarines were individually mounted with the exception of the main turbines, which were mounted on a raft; this resulted in a much quieter submarine than HMS *Dreadnought* and her *Skipjack* class sisters which were solidly mounted to the deck. Horlick noted that experience of operation of auxiliary machinery in naval marine plant had shown the: '…inherent advantages of electrically-powered auxiliaries over steam-powered auxiliaries'.[100] There were also similar advantages in air-cooled electrical motors over water-cooled motors and the accompanying small bore pipe work. The operating and maintenance difficulties experienced with the secondary machinery in the *Valiant* submarines and *Resolution* class submarines were ascertained and were designed out of the next generation of nuclear submarines, the *Swiftsure* class.

As noted in Chap. 1, streamlining a submarine reduces drag, thus increasing its submerged speed; the other benefit is reduced flow noise, making the submarine quieter and harder to detect. However, one of the major sources of noise from a submarine is self-radiated noise, hence the focus on reducing machinery noise through better design and by installing equipment on resilient mounts so noise transmitted to the pressure hull is either reduced significantly or eliminated entirely. The Royal Navy's philosophy in any engineering development can be described as seventy percent evolution and thirty percent revolution; this could be seen in the decision not to adopt all of the recommendations of the comparative study between the S5W plant and the PWR 1 plant intended for Dounreay, as discussed in Chap. 5. Some improvements were implemented in HMS *Valiant*, whilst others were left to be adopted in the *Swiftsure* class. In conjunction with the Core B development programme, a contract was placed with Yarrow Admiralty Research Department (Y-ARD) and Vickers to examine the design basis for the secondary propulsion machinery, this was delayed by lack of funding until 1964 with the start-up of the Polaris programme.

The new submarine project was designated SSNOX (*Swiftsure* class) and was to incorporate lessons learnt from HMS *Valiant* with an emphasis on operation and maintenance of the secondary machinery plant and further noise reduction measures. The Sea Water Services system for cooling HMS *Valiant*'s secondary and auxiliary machinery was not only complicated, it also meant that a lot of the pipe work in the machinery spaces was

[100] Horlick, 'Submarine Propulsion in the Royal Navy', p. 12.

exposed to deep diving depth, which added to the vulnerability of the submarine. The system supplied a number of services, including the turbo-generators and main turbines, which on a *Valiant* submarine were sited in different compartments and had separate seawater-cooled condensers. The SSNOX project designed a new machinery layout in which the turbo-generators and main turbines shared a single condenser, which, with carefully sited main circulating water pumps, considerably reduced the amount of pipework in the machinery space. This reduced the submarine's vulnerability and also allowed for better access for maintenance. The main propulsion machinery, which includes the main gearbox and the equipment mentioned above, were mounted on a single machinery raft which absorbed the machinery noise and, even at high power, could manage: '...the higher torque associated with the slower speed SSNOX propulsor'.[101] The *Swiftsure* class would, however, keep the same manoeuvring room layout as HMS *Valiant*; a new manoeuvring room layout would be introduced in the later *Trafalgar* class.

To test the prototype machinery, the MoD (N) had requested Vickers to submit alternative proposals for using services at either the Admiralty Development Establishment at Barrow (ADEB) or the British Ship Research Association (BSRA) site at Wallsend on the Tyne. The facilities at ADEB were approved and authority was sought from the Treasury to proceed with secondary propulsion machinery full power trials at an estimated cost of £700,000.[102] ADEB was nearby the shipbuilding yard and had been used for design and development of the HTP propulsion system as well as testing the DSMP secondary machinery. Upon successful completion of trials with the prototype machinery being tested at ADEB, the machinery set was to be installed into the second of the *Swiftsure* class, SSN09. The initial intention was SSN08 would be built as a repeat *Valiant* class; this was probably due to the tight building schedule, which, in the late 1960s, was launching a nuclear submarine on average every nine months. There was a requirement to fully prove the new machinery in a sea-going submarine, although prototype machinery can be run on a shore test bed, it cannot be subjected to the forces that it would be exposed to on an operational submarine such as the angles experienced when going deep or coming shallow, and "banking" when turning at speed. For reasons not discovered, SSN08 was eventually built as a *Swiftsure* class, and it

[101] Horlick, 'Submarine Propulsion in the Royal Navy', p. 12.
[102] TNA, T 225/3324, Letter, Hinxman to Truman, 27 October 1966.

became urgent that long lead items be ordered so that the programme did not suffer any grid-locks in the supply chain.[103] Horlick noted that problems did occur, such as a gearbox failure and main turbine thrust bearing failures; however, the risk was assessed and a decision was taken to fit the machinery onboard to prevent delays, any solutions forthcoming from the prototype set would be carried out in situ.[104] The secondary machinery plant in the *Swiftsure* class was a great success in regards to operational reliability and for maintenance; for the follow-on *Trafalgar* class the focus on development would be directed towards further improvements in noise reduction.

There are many other machinery and engineering developments in a nuclear submarine that are a huge improvement over conventionally powered submarines, which are not dealt with here due to the subject matter not being directly relevant to nuclear propulsion. However, one of the biggest improvements which merits a mention was in the sphere of air purification which had a direct effect on habitability and would make the most of the benefits of nuclear propulsion. Royal Navy diesel submarines in service after World War II had a crew of between sixty and seventy plus men dependent upon the class of submarine; they were limited to supplies of fresh air by use of the snort mast and surfacing. Whilst dived, the submarine's crew relied on an elementary chemical means of producing oxygen and eliminating carbon dioxide. A most basic means of producing oxygen is by use of sodium-chlorate candles; these are burnt at a high temperature in a generator and will produce a given amount of oxygen. Carbon dioxide can be cleansed from the submarine's atmosphere by use of soda-lime canisters which, when attached to an absorption unit, draw air through the soda-lime thus cleaning through chemical reaction some of the carbon dioxide. Elementary air purification indicators would measure oxygen and carbon dioxide levels, and tables would dictate how many oxygen candles to burn and carbon dioxide canisters to run given the amount of crew onboard and other factors. The amount of air purification stores that can be carried on a submarine is determined by the amount of space available and is another factor in limiting dived endurance. Initially, HMS *Dreadnought* would carry an amount of these air purification stores for everyday use until other, more effective methods became available; these basic chemical stores would also be carried for emergency use and

[103] TNA, T 225/3324, Letter, N/S 1004/67, 24 February 1967.
[104] Horlick, 'Submarine Propulsion in the Royal Navy', p. 13.

are stowed in the submarine's escape compartments located in the forward and after compartments. Although the nuclear-powered submarine is larger than its conventionally powered counterpart, space is still at a premium and there is a larger crew of over a hundred when carrying trainees all living in a sealed environment for days, weeks and sometimes months on end. The use of chemical stores is not an efficient means of controlling a submarine's atmosphere as the stores carried take up valuable space. However, a nuclear power plant produces enough electrical power to consider other methods of submarine air purification; these were being investigated in the 1950s alongside the development of nuclear propulsion.

Many people understand the basic principles of electrolysis whereby a direct electrical current is passed through water and oxygen is collected at the anode and hydrogen at the cathode; it was this process that was being researched for submarine use. A report was produced by the Admiralty which detailed what equipment was available in the mid-1950s, their operating parameters, their power usage and their dimensions. The report concluded that electrolysis of water under pressure would provide a convenient, compact and safe method of obtaining oxygen for breathing purposes.[105] The high pressure electrolyser was adopted as the unit for production of oxygen which is passed through to the submarine's ventilation system, the hydrogen is compressed and discharged overboard. A basic CO_2 Scrubber removes carbon dioxide from the submarine's atmosphere by passing it over a bed of amine, a chemical which absorbs carbon dioxide when cold; as it is heated the amine and carbon dioxide are separated and the carbon dioxide is compressed prior to discharge overboard. CO/H_2 Burners use a catalyst bed to turn the carbon monoxide into carbon dioxide which is then dealt with by the CO_2 Scrubber, the hydrogen is turned into water vapour. Other environmental apparatus deal with hydrocarbons, odours and particles; many articles taken for granted in every household such as aerosols, paints and shoe polish are either prohibited or strictly controlled for onboard use thereby eliminating possible sources that may have to be dealt with by the submarine's air purification system. Indeed, radiation monitors onboard nuclear-powered submarines are so sensitive that certain older styles of watches with "illuminated" faces are also banned from onboard use. Finally, an atmospheric analyser constantly monitors the submarine's atmosphere for concentrations of the

[105] TNA, ADM 287/40, Electrolytic oxygen production for breathing on nuclear powered submarines, 1957, Report, DEMR/ER3/30, May 1957.

gases noted above as well as nitrogen and Freon. These items of air purification machinery allow the crew to take full advantage of a nuclear-propelled submarine's ability to operate independently from the atmosphere.

Reporting the Nuclear Navy

Before the cultural changes of the 1960s, when the "Establishment" and authority in general began to be distrusted and challenged, even satirised, reports in the press and other media followed the conventions of the day reporting facts and informing the general public. There was no real analysis or informed expert opinion which one would expect in today's media. As such, during the 1950s and early 1960s, although a few details on the Admiralty's nuclear submarine programme were reported in the press, these were mainly confined to debates in Parliament. Typical of such factual reporting was an article on the press being invited to see the Jason reactor at the Royal Naval College, Greenwich, as described in Chap. 5. There was no further comment about it being located in an iconic Grade 1 listed building nor of its siting within Greenwich; research failed to uncover any correspondence relating to this decision in the local or national press.

The Admiralty did not hold any public relations event for the nuclear submarine project when approved by the Treasury in January 1956; however, reports did start to appear in the national press later that year commenting on the LIDO reactor at Harwell becoming critical and starting shielding experiments for the nuclear submarine programme.[106] Further reports appeared in *The Times* and other papers throughout 1957 advising that good progress was being made with the design of the British pressurised water reactor and of various contracts being awarded. During 1958, as negotiations were taking place with the US, reports appeared of Rickover's visit to the Admiralty and of the eventual purchase of the S5W plant as part of the Mutual Defence Agreement but little else was forthcoming. It was only in 1959 that other media interest began to take notice. The unusually high-profile event of Prince Philip performing a keel-laying ceremony for HMS *Dreadnought* was not only reported in *The Times* and on BBC radio's Home Service channel, it was also recorded by British

[106] 'Atomic Reactor Starts Work', *The Times*, 22 September 1956, p. 6, Issue 53643.

Pathé and distributed as newsreels to cinemas across the country.[107] A shorter news cast was released by Reuters, possibly for overseas markets; in the days when few households had television, these newsreels would have ensured that a mass audience would have been informed about Britain's nuclear submarine programme.

The building and fitting out of HMS *Dreadnought* was marked by a number of walk-outs and strikes by various Trades Unions, some were for a short period whilst others lasted weeks; these were reported in national and local press and indicate that all aspects of the Admiralty's programme were covered; a dispute involving the Boiler Makers Society prompted the headline: 'Atomic submarine may be delayed'.[108] On 10 July 1962, both the *Birmingham Daily Post* and the *Newcastle Evening Chronicle* reported on another walk-out at Barrow of 128 electricians involved in a demarcation dispute. Throughout the UK the local press was as active as the national press in reporting on the progress of HMS *Dreadnought*; the *Aberdeen Evening Express* wrote an article on the imminent conclusion of the purchase of the S5W reactor plant, 23 February 1959. An article appeared in the *Liverpool Echo*, 12 November 1962, noting that HMS *Dreadnought*'s reactor had gone critical for the first time, and a couple of weeks later the *Belfast Telegraph* reported on HMS *Dreadnought* sailing from Barrow to Rothesay for sea trials, 26 January 1963. However, reports were not limited to the submarine; the crew was also subjected to media attention. A report in March 1959 announced the name of the Commanding Officer designate of HMS *Dreadnought*, Lieutenant Commander B. F. P. Samborne (and of Hammersley as Engineering Officer); the report also noted that sixteen officers and thirteen senior rates had been selected for nuclear training.[109] On 23 April 1959, the *East Kent Gazette* reported that four officers and eight ratings were going to the US for nuclear sea training. The *Coventry Evening Telegraph* reported 10 June 1960 that HMS *Dreadnought* would carry its own medical staff, something that other submarines did not do. With the advent of Polaris, a new reality had to be confronted by all crewmen, whether or not to receive bad news on patrol. Being unable to respond to a bereavement or other

[107] 'Saucer and Science – Barrow in Furness', British Pathe, 15 June 1959, Film ID 1585.21.

[108] 'Atomic Submarine may be Delayed', *The Times*, 30 June 1959, p. 5, Issue 54500.

[109] 'Atomic Submarine Commanding Officer Named', *The Times*, 25 March 1959, p. 10, Issue 54418.

such domestic crisis whilst on patrol, the *Daily Mirror* reported on this heartbreak choice for families in its 28 June 1969 edition. It is apparent from these reports that the Admiralty was providing information to the media in order to raise public interest in all aspects of the nuclear submarine programme. Some reports, however, were not reliant on the Admiralty providing the material.

In the late 1950s, the Campaign for Nuclear Disarmament (CND) came to prominence, demonstrating against the UK's pursuit of nuclear weapons. From the CND files researched at the London School of Economics, little was found on the subject of nuclear submarines; only the written questions were posed to the Defence Minister, Denis Healey, 16 September 1964, concerning the four Polaris submarines. It is possible that because of the apparent "lack of interest" in nuclear submarines by CND during this period and, indeed, the UK's embarrassing security issues during the 1950s and early 1960s (Klaus Fuchs, the Cambridge Spy Ring and the Portland Spy Ring, to name the most infamous), the physical security of nuclear submarines does not appear to have been taken as seriously as information security by the Admiralty. There were security breaches, and these were reported upon. *The Times* and the *Belfast Telegraph* both reported in early March 1962 on an inquiry into how the word "Rag" had been painted (probably by students) on the side of HMS *Dreadnought* whilst at the Devonshire Dock, Barrow. Later that year, it was reported in the *Birmingham Daily Post* that CND protestors had managed to gain access to the casing of the USS *Nautilus* whilst at Portland, 16 July 1962, the protestors were later released by police without charge. Although these incidents would have been embarrassing to the Admiralty, they do show that, whilst not questioning the apparent lack of security in their articles, the press were willing to report on any story connected to nuclear submarines.

The launch of HMS *Dreadnought* by the Queen in 1960 was a very important moment for the Royal Navy, and one could argue for the country; as such the ceremony was broadcast by BBC television on the morning of the launch and recorded for the BBC Home Service so households without television could also listen to the event. After the launch of HMS *Dreadnought*, nuclear submarines increasingly became a topic of debate on radio and on television. A few days later, capturing the public's interest, the BBC broadcast a programme detailing the advances made in submarines, through personal reminisces, since the first submarine, *Holland 1*, commissioned in 1901.[110] HMS *Dreadnought*'s commissioning ceremony on 17 April 1963 was also widely reported throughout the UK. In the US, there appear to have been no reports in the *Washington Post* nor the *New York Times*, not even around the Groton area of Connecticut, with its strong connection to *Dreadnought*. This was possibly due to the fact that the USS *Thresher* was lost at sea 10 April 1963, and American focus was inevitably on the 129 men who lost their lives and on investigating how such a disaster had occurred. However, the commissioning was reported later throughout the US in places such as Kansas and Michigan.[111]

In the UK, throughout 1963 there were steady press reports and newsreel films of various aspects of HMS *Dreadnought*'s sea trials and of the launch of the Royal Navy's second nuclear-powered submarine, HMS *Valiant*, which was performed by "Mrs Peter Thorneycroft", the Defence Minister's wife, as recorded by the newsreel.[112] Reports about nuclear submarines continued through the 1960s, but these articles started looking at the complications that began to appear, such as the Inconel problem at Dounreay, which was reported in *The Times*. The hairline cracks found in some of HMS *Dreadnought*'s steel plates, were reported in both *The Times* and the *Daily Mirror* as was the subsequent decision to buy US HY80 steel because UK producers could not supply the right quality of material. Interestingly, however, there was none of the sensationalism that is associated with today's media; indeed, when HMS *Dreadnought* was in dock for investigation of the cracks in the steel plates, it was reported that a fire broke out in the control room but: 'It was soon put out by the crew'.[113] Despite these reports, articles still appeared which catered to the public interest; the BBC produced programmes provoking debate on naval power and of the nuclear deterrent. After the signing of the Nassau Agreement in December 1962, BBC television broadcast an NBC special 22 January 1963, featuring the USS *George Washington*, the US's first Polaris-armed submarine. The programme gave the UK public an insight into the future national deterrent based on Polaris. Many reports and programmes were produced with the assistance of the MoD (N); the BBC Home Service broadcast a programme 28 August 1966, 'Down Your Way', featuring

[110] 'White Mice to Dreadnought', BBC Home Service, 25 October 1960 19:30. www.genome.ch.bbc.co.uk

[111] For example, the *Arkansas City Traveler*, Kansas, 18 May 1963 and the *Ironwood Daily Globe*, Michigan, 6 May 1963. www.newspaperarchive.com

[112] 'Britain's Second Atom Submarine', British Pathe, 5 December 1963, Film ID 1780.29.

[113] 'News in Brief', *The Times*, 21 December 1965, p. 5, Issue 56510.

HMS *Valiant*. Another submarine, HMS *Resolution*, was the title and featured in a Radio 4 broadcast 7 May 1969. It must be apparent that gaining access to an operational submarine to produce a programme or research an article would require a large amount of planning.

Members of the media attempting to visit a submarine would have been involved in months of negotiations and possible background checks before being allowed to set foot onboard. The Admiralty, and later the MoD (N), would have requested a summary of their intentions and would have sought some editorial control over what was allowed to be reported/broadcast. The articles and programmes were by no means sterile reports but features such as submarine and machinery operating parameters would have been strictly off limits. Filming in certain areas would have been prohibited and in areas where approved, the camera shot would have been posed so as not to capture instrumentation and equipment in the background. No doubt the Admiralty saw the reporting of nuclear submarines in a positive light and did use them for recruiting purposes; the numerous firms involved in constructing nuclear submarines also used them as advertising for apprentices. Where firms were also involved with supplying industry, they too used their participation in nuclear submarines to project the image of the company being at the forefront of technology. The Royal Navy continues to this day to allow limited media access to submarines; this openness with the media is the antithesis of the submarine branch, widely known as the "Silent Service". It was, and still is, a difficult line to balance with operational security being paramount, but it is because public and media access is strictly limited that submarines continue to maintain their mystique.

Project Cost

A memorandum prepared for the DRPC gave a total through costing of £32.16 M from 1958 to 1965 for HMS *Dreadnought* and the Dounreay prototype.[114] The First Lord broke down these figures for the Prime Minister noting £18.5 M was projected to be spent at Dounreay and £8.85 M on purchasing the S5W reactor and its associated machinery from the US.[115] The last record researched of the Dounreay prototype estimates noted that the £26.5 M held firm but was dependent on

[114] TNA, DEFE 69/749, Minute, D(58)(72), 25 November 1958.
[115] TNA, PREM 11/2635, Letter, Selkirk to Macmillan, 27 November 1958.

reaching criticality in September 1963; however due to the Inconel problems encountered and the costs incurred in rectifying them, criticality was not achieved until January 1965.[116] The Naval Estimates for 1963–1964 indicate a build cost figure for *Dreadnought* of £18 M, it is fair to assume that the whole nuclear propulsion project had cost in the region of £40–50 M. It begs the question that, apart from the development cost, exactly what did the Government get for its money? Excluding guns and certain other equipment, the following build costs are taken from the same Naval Estimates for comparison: HMS *Dreadnought* £18,055,000, the guided missile destroyer HMS *Devonshire* £14,080,000, the *Leander* class frigate HMS *Leander* £4,630,000 and the *Oberon* class submarine HMS *Olympus* £2,500,000.[117]

Cost comparisons between different classes of warships and submarines do not take account of the dissimilar operations they undertake; therefore, it is difficult to assess a vessel's "value" in relation to other platforms. The warship can be used in multiple roles away from its primary function such as reaction to natural disasters and humanitarian missions; a warship's presence can impose economic sanctions, such as those against Southern Rhodesia discussed in the introduction, or even engage in limited diplomatic disputes such as the Icelandic "Cod Wars" of the early 1970s. The warship can engage in limited hostilities and if required escalate the fighting. However, the submarine is ill-suited to most of these roles and can only be employed to threaten, as encapsulated by the role of the deterrent SSBN submarines or commit acts of violence against an enemy. It could be argued that seven conventionally powered submarines could be purchased for the price of one nuclear-powered submarine; however, many would contest that this would be a false economy. This book has shown that the nuclear-powered submarine is not limited in terms of endurance and capability as is its conventionally powered counterpart; its "value" in the role of the national deterrent cannot be sufficiently expressed; the threat of using nuclear weapons to prevent major war is certainly controversial but its success in preventing, since World War II, another major global war cannot be denied.

It is implicit from this chapter that with core improvement comes improved safety, simplicity of operation and ease of maintenance. The safety and reliability of the PWR 1 reactor plant, and its successor PWR 2,

[116] TNA, T 225/2206, Letter, Jones to Fraser, 1 July 1963.

[117] Parliamentary paper, Navy Estimates 1963–64; 1962–63 (113), p. 70.

can be realised by the fact that the Continuous at Sea Deterrent (CAS-D) of the Polaris- and Trident-armed submarine patrols has been maintained for over fifty years. It is one thing to have a safe and reliable nuclear power plant; however, full advantage can only be grasped if its secondary machinery is as reliable. The Royal Navy was producing a steam raising plant (the pressurised water reactor) just as it was designing its surface warships to be powered by gas turbine and it has been seen that initially not enough design effort was accorded to the secondary machinery as it was deemed a proven technology. All good engineering lessons learn from past mistakes, new ideas are taken forward and good design is improved upon. The electrical generating power of a nuclear power plant allowed for the introduction of new technologies, especially those in the field of air purification, which allowed the crew to take full advantage of the nuclear power plant's independence from air.

The striking improvements in submarine technologies can best be characterised by Vice Admiral Sir Arthur R. Hezlet CB DSO DSC (Flag Officer Submarines 1959–1961), who wrote: 'If the old fashioned U-boats overall submerged performance is represented by m, then the Type XXI would be 10m, the Walther boat 40m and the nuclear submarine at 5,000m'.[118] HMS *Trenchant*, the oldest of the *Trafalgar* class submarines still in operation was commissioned 14 January 1989; two other 'T' class submarines are also still in commission, HMS *Talent* and HMS *Triumph*.[119] That is a testament to the core designers, the build quality and the performance of the PWR 1 reactor, including its associated secondary machinery. It is also testament to the Royal Navy marine engineers, artificers and mechanics, assisted by dockyard engineers, whose skill and professionalism keep a nuclear-powered submarine operational for thirty or more years.

[118] Hezlet, *The Submarine and Sea Power*, p. 244.

[119] Note: Hezlet was the Commanding Officer of the World War II submarine *Trenchant*, in which he won a Bar to his DSO. Hezlet launched the new *Trenchant* and kept in regular contact with its crew.

CHAPTER 8

Conclusions

During a debate in the House of Lords on the Naval Estimates concerning the possible means of submarine nuclear propulsion, Viscount Hall, First Lord of the Admiralty 1946–1951, noted: 'Some United States Navy planners go as far to say that, just as the carrier superseded the battleship, so might the Dreadnought of the deep, the future submarine, become the capital ship of the future'.[1] It was prescient that Hall evoked the name *Dreadnought* during the debate. In naming its new, nuclear-powered submarines, the Royal Navy eschewed previous famous submarine names such as HMS *Upholder* and HMS *Seraph*, and gave its new submarines the names of famous battleships which reflected the Royal Navy's opinions on the status of these vessels.[2] Beginning with HMS *Dreadnought*, the battleship design that revolutionised naval warfare in 1901, HMS *Valiant* and HMS *Warspite* soon followed with the Polaris submarines carrying the names of *Resolution, Renown, Repulse* and *Revenge*. The nuclear-powered submarine is the "hunter-killer" envisaged by Mountbatten and the carrier

[1] Parliamentary debate, House of Lords sitting 27 March 1952, 1951–52; Elizabeth II year 1; Columns 1023–1098, Fifth Series Volume 175.

[2] See Duncan Redford, 'The 'Hallmark of a First-Class Navy': The Nuclear-Powered Submarine in the Royal Navy 1960–77', *Contemporary British History*, Vol. 23, No. 2, (Jun., 2009), 181–197.

G. M. Jones, *The Development of Nuclear Propulsion in the Royal Navy, 1946–1975*, Security, Conflict and Cooperation in the Contemporary World,
https://doi.org/10.1007/978-3-031-05129-6_8

of the national deterrent. The historians, Peter Hennessy and James Jinks, wrote of Royal Navy submarines: 'Their world spans the front line of national defence (surveillance and intelligence gathering) to the last line (nuclear retaliation as the country's near unthinkable 'last resort')'.[3] The nuclear-powered submarine is the capital ship of today's Royal Navy, all this has been made possible with the successful development of nuclear propulsion. However, as discussed with evidence provided from files at the National Archives and elsewhere, at its inception the nuclear propulsion programme suffered from low priorities, indifferent management and inadequate resources.

Political Problems and Indifference

An expert on Anglo-American relations, John Baylis has argued that during the period the McMahon Act was in force there was an "epistemic community" of military personnel, intelligence officers, government officials and nuclear scientists on both sides of the Atlantic who maintained a close working relationship: '...which was of crucial importance in breaking down the barriers to collaboration which were erected after World War II'.[4] Such examples have been presented; in Chap. 3 it was noted how, during the technical team visit to the US in 1957, a piece of reactor-grade hafnium came into their possession once it was known that they did not have any for testing. In Chap. 4 it was noted that Rickover made a trans-Atlantic call to a British metallurgist (probably the Harwell metallurgist Finniston) to clarify a technical point.[5] These examples indicate that, whilst comprehensive data was not exchanged, low-level nuclear information was exchanged to the benefit of both parties. Politically, however, full exchange of nuclear information would have to wait until Eisenhower replaced Truman as US president and Macmillan became prime minister; both men had enjoyed a close working relationship during World War II. Soviet success with the launch of *Sputnik*, which alarmed the US, was

[3] Peter Hennessy & James Jinks, *The Silent Deep: The Royal Navy Submarine Service since 1945* (London, Allen Lane, 2015), p. xxx.

[4] Baylis, 'The 1958 Anglo-American Mutual Defence Agreement:', p. 426.

[5] Hewlett and Duncan, *Nuclear Navy*, p. 246.

the catalyst to the repeal of the McMahon Act; Botti, who has written on Anglo-American relations, observed: '...the Eisenhower administration accomplished more in Anglo-American relations in two months than American officials had in the previous twelve years'.[6]

It is evident from the files that the nuclear submarine project, whilst under political scrutiny by the First Lords of the Admiralty, only became of interest to politicians when the US made the offer for the UK to purchase a submarine propulsion plant. Even then, the purchase was viewed as a possible means of reducing expenditure on the project. From 1946 onwards, it is fair to view this period as one of political indifference. It was only in the early 1960s, however, with the cancellation of the stand-off Skybolt missile system and the signing of the Nassau Agreement and later the Polaris Sales Agreement, that British politicians grasped the potential of what nuclear propulsion offered—a sea-based deterrent. Macmillan advised the Queen that his philosophy, for political and moral reasons, was to rid the UK of land-based rockets; the best thing was to move the deterrent out to sea where it would be more acceptable: '...in a submarine, is out of sight'.[7] Political use of the nuclear-powered submarine would continue to develop as the expanding submarine activity of the Soviet fleet increased the need for vigilance, and new threats to UK security emerged.

Technical Considerations

The level to which engineering standards needed to be raised by all groups involved in the nuclear propulsion programme has been identified and presented; however, it can be argued that it was not the purchase of the S5W reactor technology per se that was of most benefit to the programme, but the introduction of US standards and techniques. Many of the officers interviewed have referred to the US nuclear engineering philosophy and the establishment of innovations such as the rip-out and tag-out systems for controlling system and machinery configurations which helped to secure nuclear propulsion in the Royal Navy.[8] Captain Farley-Sutton's

[6] Botti, *The Long Wait: The Forging of the Anglo-American Nuclear Alliance 1945–1958*, p. 164.

[7] Horne, *Macmillan 1957–1986: Volume II*, p. 276.

[8] Note: "tag-out" is a physical notice used to indicate that a valve or equipment has been isolated and not be operated due to maintenance or other purpose. The "rip-out" is a procedure for controlling the tag-outs in place and the authorisation for their removal and the valve or equipment returned to safe operation.

views are recorded in Chap. 4; Rear Admiral Hammersley has also advised: '...that one of the greatest benefits was adopting the philosophy of test groups and safety management'.[9] This was also the view of Professor Jack Edwards who wrote: '...that the US deal embraced standards of controls in quality, procedures and safety that we were only just beginning to appreciate as being absolutely essential to ultimate success'.[10] The Principal Admiralty Overseer during HMS *Dreadnought*'s construction, E. A. Brokensha, wrote in a letter containing lessons learnt during her build that the institution of test forms in HMS *Dreadnought* was: '...a significant factor in producing very satisfactory sea trials and a Test Form procedure is recommended for all ships building'.[11] The organisation for developing a nuclear propulsion plant was dispersed and only became a coherent whole on the formation of the Dreadnought Project Team with the Technical Chief Executive, Rowland Baker, controlling all aspects of the project and the team based at one location, Bath. It was this team and its adoption of US standards and practices which they disseminated to other firms and parties involved in the project that made the development of nuclear propulsion in the UK a success.

The Ministry of Supply and its successor, the UKAEA, in authorising the work carried out at the AERE Harwell, would have had a responsibility to account for money it spent on research and development and to recover costs wherever possible. With limited enriched fuel available in the UK and hazards identified in the design of the Hanford water-cooled and graphite-moderated reactor in the US, the UK concentrated its research on gas-cooled and graphite-moderated reactors. The experience gained at Calder Hall had indicated that by building power stations with larger outputs it should be possible to produce electricity more cheaply than at Calder Hall.[12] These were viewed as having the potential to compete with coal-fired generating stations and a means of retrieving some of the costs incurred in development. This concentration of effort on the design of gas-cooled and graphite-moderated reactors to the detriment of water-cooled and water-moderated reactors certainly delayed development of

[9] Private correspondence, Rear Admiral Peter Hammersley, 29 June 2018.

[10] Private correspondence, Jack Edwards to Vice Admiral Sir Robert Hill, 17 April 1998.

[11] TNA, ADM 1/29274, HMS/m Dreadnought (SSN 01): Lessons learnt during construction, 1963–64, Letter, Brokensha to Smithers, 26 May 1964.

[12] Jukes, J. A., 'Nuclear Energy: A Survey of Britain's Position', *International Affairs (Royal Institute of International Affairs 1944–)*, Vol. 32, No. 3, (Jul., 1956), 273–282 (p. 275).

the Royal Navy's first nuclear-powered submarine. Indeed, as noted in Chap. 2, Harwell's chief engineer, H. Tongue, was against research into light-water reactors due to their limited thermal efficiency, which would restrict further development. Therefore, it was not only the lack of enriched uranium that was a barrier to developing the pressurised water reactor, the cost to the UKAEA's budget of developing a reactor that was perceived to have limited potential for further development was also a determining factor and, hence, a restriction to the Royal Navy's nuclear submarine programme.

The Military Situation

Attention has been given to the challenges the Royal Navy faced in developing nuclear propulsion, and this book has explored the reasons for requiring submarines with a high endurance; however, the threats and challenges the Royal Navy faced were fluid throughout the Cold War. At the start of the project, the main maritime concern to the UK was the threat posed by the "fast submarine". As discussed in Chap. 1, this threat initially revealed itself in the submarine developments of World War II (Fast Battery Drive and High Test Peroxide submarines), which the UK and US knew the Soviet Union would fully exploit. Stalin was already planning the creation of the world's most powerful submarine fleet: '...using captured German models and scientists to that end'.[13] Given the UK's reliance on maritime trade and the near success of German U-boats during the two World Wars, the Royal Navy sought to counter this threat by making the detection and destruction of enemy submarines the priority of its submarine flotilla.

The UK's proximity to the Barents Sea, into which the submarines of the Soviet Northern Fleet sail from their bases on the Kola Peninsula; the UK's bases in the Mediterranean; and the Royal Navy's ability to patrol the Greenland–Iceland–UK (GIUK) gap, all natural choke points for Soviet submarines transiting to the US eastern seaboard, amounted to less transit time and a longer period in the patrol area for Royal Navy submarines than their US counterparts. This would be of great benefit to the US and it would certainly make military sense for the Royal Navy to assist in

[13] C. Jay Smith and Ernest J. King, 'The World War II Soviet Navy Revisited: A 1965 evaluation and a look at the future', *Naval War College Review*, Vol. 18, No. 7, (Mar., 1966), 10–51 (p. 42).

detecting Soviet submarines as they were leaving their bases, especially high-priority targets such as ballistic and cruise missile submarines. The authors, J. J. Tall and Paul Kemp explored this possibility arguing that it was not hard to see why the US would want its traditional ally to join the "nuclear propulsion" club, especially by 1967 with the introduction of the Soviet *Yankee* class (SSBN), *Charlie* Class (SSGN) and *Victor* class (SSN): '...all of which added immense power and flexibility to the Soviet submarine fleet since they represented a step change in technology'.[14] It has been shown that the datum given by the purchase of the S5W reactor plant, the improvements brought about by core developments, machinery rafting and other engineering innovations allowed Royal Navy submarines to maintain their technological advantage over the Soviet opposition throughout the period discussed.

The Nuclear Propulsion Legacy

There is one legacy that cannot be ignored and that is the waste that has been produced and stored since the inception of the Naval Nuclear Propulsion Programme. Since 1965, Dounreay has operated five different submarine reactor cores: cores A, B and Z in the original PWR 1 and cores G and H in the Shore Test Facility (PWR 2), which finally shut down 21 July 2015. The process of decommissioning and disposal of both prototype reactors is ongoing with the site being cleared of fuel elements.[15] The disposal of sixteen SSNs and four SSBNs has been more problematic; concern for the environment and the geographic limitations of the British Isles have added to the impetus to find an effective solution. All submarines decommissioned are currently laid up waiting disposal; the four Polaris SSBNs and three SSNs are laid up at Rosyth Dockyard and the remaining thirteen SSNs are laid up at Devonport Royal Dockyard's 3 Basin (*Courageous* is open as a museum ship). The majority of these submarines have had their cores removed but a number are waiting for dock availability, following work to improve safety standards, to allow docking and, thus, defueling to proceed. However, decommissioning is progressing; *Swiftsure* has recently completed a successful docking period at Rosyth

[14] J. J. Tall & Paul Kemp, *HM Submarines in camera: An illustrated history of British submarines, 1901–1996* (Leicester, Blitz Editions, 1998), p. 173.

[15] https://www.gov.uk/government/news/shutdown-of-shore-test-facility-reactor [Accessed 21 February 2019].

dockyard to remove low-level waste from the reactor compartment, and *Resolution* is currently undergoing the same process (Fig. 8.1).

From the purchase of the S5W plant, the Royal Navy's legacy has been one of continuous improvement in core endurance and power which has led, and will lead, to considerable savings in through-life costs for a submarine. Core A, which was based on the S5W core, went critical in January 1965 and completed its burn-up at Dounreay in 1967; the improved Core B went critical in August 1968 and completed its burn-up in early 1972.[16] The final core improvement for PWR 1, Core Z, initially went critical at Dounreay in December 1974 and following resolution of some technical difficulties began its run in June 1976. Core Z completed its burn-up in June 1984, which indicates just how much endurance the core designers had built into the PWR 1.[17] With PWR 1 at the limits of its development potential, a new PWR 2 was designed and Core G first went critical in July 1987 completing its burn-up in December 1996; from a nuclear-powered

Fig. 8.1 HMS/m *Repulse* alongside awaiting trials 05/10/1968

[16] Lambert, *Rolls-Royce: the Nuclear Power Connection*, p. 56.
[17] Lambert, *Rolls-Royce: the Nuclear Power Connection*, pp. 64–65.

submarine requiring a refit to refuel every three to four years, the submarine would require only one refit to refuel during its lifetime.[18] The *Astute* class submarines, fitted with the improved Core H, will never require refuelling during their predicted thirty-plus-year operational life; the savings in cost and to the environment (in the form of used fuel elements) are significant. Furthermore, with regards to protecting the environment, the new generation of PWR designed and built for the new *Dreadnought* class SSBN (PWR 3) will not require a shore prototype as this work is now done by computer modelling. With the requirement to refuel eliminated, platform availability is also improved, as a submarine will no longer be required to dock for periods of three to four years; it could be argued that this increased availability will require fewer submarines to accomplish the operational commitments dictated by government.

Arguably, the enduring legacy of nuclear propulsion has been the Royal Navy's ability to maintain the UK's Continuous At Sea Deterrent (CAS-D) through the Polaris- and Trident-armed submarines. HMS *Resolution* sailed on the first Polaris patrol 15 June 1968; she was joined in the task by HMS *Repulse* in the spring of 1969. On 14 June 1969, the Royal Navy's Polaris submarines formally took over the primary deterrent role from the RAF's V bomber force.[19] The year 2019 marked fifty years of continuous sea-based deterrence, achieved primarily through the successful introduction and development of nuclear propulsion, in part due to the generosity of US assistance in selling the S5W plant to the UK. It is, without doubt, the politicians that gained most from this legacy despite their initial lack of involvement. Without the timely development of nuclear propulsion, I would argue that it is very doubtful that the UK would still be a nuclear armed state. Ballistic missiles are large weapons and require a submarine with a displacement and enough generating power to accommodate them. Without its nuclear deterrence it is equally doubtful that the UK would have kept its permanent seat at the UN Security Council and could continue to exert its diplomatic and political influence on its major ally, the US. As Ernest Bevin is quoted as saying: '…I don't want any other Foreign Secretary of this country to be talked to or at by a Secretary of State in the United States as I have just had in my discussions with Mr. Byrnes'.[20] It is evident since the signing of the Mutual Defence

[18] Lambert, *Rolls-Royce: the Nuclear Power Connection*, p. 83.
[19] Hennessy, *Cabinets and The Bomb*, p. 14.
[20] Hennessy, *Cabinets and The Bomb*, p. 48.

Agreement that the US has trusted the UK's diplomatic advice and military cooperation in a number of foreign policy areas; conversely, the UK has also on occasions used its own policies to act independently of US wishes (such as involvement in the Vietnam War).

Research in the "Secret (Nuclear) State"

The "Secret State" is a rather nebulous term with its connotations of dark and sinister actions outside the control of Parliament; a definition more suited to conspiracy theorists. In the UK, most people would associate the term with government intelligence services, MI5, MI6 and GCHQ; the more erudite would include the "Whitehall Mandarins" of the Civil Service or as Roger Verrier described them: '...the Permanent Government'.[21] However, there are other areas where government departments do not court publicity which could be included in a definition of the Secret State; anything to do with nuclear matters would naturally fall into this category as would parts of the Armed Forces such as the Submarine Service (on which governments rarely comment). Nuclear matters and submarines constitute the topic of this research, and examination into the subject has shone light into an area where little academic work has been done. Research of this kind is helped where privileged access to unreleased files can be granted; also of great service are introductions to personnel involved in the subject who are willing to assist in the research; clearly, it is difficult to arrange this type of access. Government departments and other bodies engaged in nuclear matters are very guarded in their dealings with members of the public who purport to be interested in their work for academic or other purposes. Indeed, overtures were made requesting access to Rolls-Royce's archives; these were declined but in reply I did receive a copy of Lambert's book, *Rolls-Royce: the Nuclear Power Connection*. A great deal of trust is placed in the researcher who is given access to unreleased documents to respect the original security caveat and be sensitive to the information contained therein. As Arnold noted, access to unreleased files does not give the historian permission to publish but they do add an extra dimension and deeper understanding to their research.[22]

[21] Quoted in Peter Hennessy, *Whitehall* (London, Secker and Warburg, 1989), p. xvii.
[22] Arnold, 'A letter from Oxford', p. 214.

It was noted in the Introduction sub-chapter of *Nuclear Historiography* that due to the secrecy surrounding the subject matter and possibly due to the controversy surrounding nuclear issues, not many people have tackled the subject. Professor Margaret Gowing and Lorna Arnold were employees of the UKAEA; Katherine Pyne was an employee of the Atomic Weapons Establishment (AWE). Gowing was the official historian for the UKAEA and was later assisted in her work by Arnold; Pyne became the official historian of the AWE after collaborating with Arnold on *Britain and the H-Bomb* (2001). Pyne also worked with Professor Lord Peter Hennessy on *Cabinets and The Bomb* (2007); these women already had security clearance to view many of the documents they would have researched as part of the Official Histories of their respective departments. Access to files held at the National Archives are open for any interested person to view, so it is entirely possible that research and writing of nuclear history in the "Secret State" could be undertaken by anyone with an interest in the subject. Indeed, Hennessy started his career in journalism and, arguably, through his reputation as a constitutional historian, became a person who has been entrusted by governments to write about the "Secret State" and has enjoyed privileged access for his research.

However, it has to be noted that in late 2018, all the United Kingdom Atomic Energy Authority files (including those that were viewed as part of the research into nuclear propulsion) held at the National Archives have been temporarily withdrawn by the Nuclear Decommissioning Authority as part of a review with the MoD and the AWE. This temporary withdrawal of files underscores the fact that nuclear security is an ongoing matter prompting periodic review. Official Histories cannot be written without sponsorship from the inside. Although research for this book has been supported by Commodore Mark Adams RN, it is not an Official History. Evidence has been presented which gives an insight into the decisions that had to be made for the Royal Navy to introduce nuclear propulsion to its submarine fleet. This book challenges traditional areas of writing about submarines; nuclear propulsion is the continuing source of fascination that inspires other authors and historians to focus on submarine operations. However, this book positions the steam raising plant, the pressurised water reactor, at the centre of the narrative. It is the reactor that is at the heart of the nuclear-powered submarine and the reason for its submerged endurance and speed. The book is new to the historiography of ship propulsion because it offers an insight into the development of nuclear propulsion in the Royal Navy and will hopefully act as a spur to further research into submarine propulsion systems. The book is an informed narrative of events from the first submission by Rear Admiral Charles Daniel in 1946 deliberating the possibility of nuclear propulsion to the successful start-up of Core Z, the final core improvement of PWR 1 in June 1976.

It has been demonstrated that the legacy of the political, military and engineering decisions, made over sixty years ago, have had a major impact on British political life and on military planning. These decisions continue to influence modern-day political and military thinkers, not only here in the UK and the US, but also among future possible belligerents. The "Successor Project" (to replace the *Vanguard* class SSBNs) is being realised; the new HMS *Dreadnought* is building and the follow-on submarines of the class have been named HMS *Warspite*, HMS *Valiant* and HMS *King George VI*. Another three submarines of the *Astute* class are either in build or have been ordered. In giving the Royal Navy a head-start with the purchase of the S5W plant, Rickover's assessment that the UK would be technically competent to produce (and improve) nuclear propulsion plants independently of the US has been justified.[23]

As noted previously, Article III of the Mutual Defence Agreement authorised the sale of the S5W reactor and propulsion plant and information relating to it. Replacement cores or fuel elements were limited to ten years, subject to terms and conditions. Article II concerns the exchange of information; paragraph five, particularly, refers to information on the research, development and design of military reactors. However, in practice, the Westinghouse/Rolls-Royce Contract, signed 10 February 1959, constituted the means by which the transfer of material and information took place. This contract was set to expire twelve months after HMS *Dreadnought*'s reactor went critical, achieved 12 November 1962. After November 1963, no further information on nuclear propulsion reactors was exchanged despite the intentions of Article II paragraph five discussed in Chap. 4. However, since the late 1990s, the US Navy and Royal Navy have collaborated on nuclear propulsion information and technology; each navy seconds a senior naval engineering officer to their respective

[23] TNA, DEFE 69/749, Minute, Selkirk to Sandys, 24 January 1958.

departments, Director Nuclear Propulsion in the UK and the Office of Naval Reactors in the US. The PWR 3 reactor is a product of that collaboration and contains elements of the S9G reactor design which is fitted to the US Navy's latest *Virginia* class SSN.[24] As a result of this collaboration, the Royal Navy will continue operating nuclear-propelled submarines well into the second half of the twenty-first century and in all probability, beyond.

[24] Turner, Julian, *Deep impact: inside the UK's new Successor-Class Nuclear Submarine* 29 July 2013. https://www.naval-technology.com/features/feature-nuclear-submarine-successor-uk-royal-navy [Accessed 2 February 2019].

Appendix A

Royal Navy Nuclear-Powered Submarines

Ship Submersible Nuclear—SSN		
HMS *Dreadnought*:	Commissioned 17/04/63	4000 tons dived (S5W Fitted)
HMS *Valiant*:	Commissioned 18/07/66	4500 tons dived (PWR 1 Core A)
HMS *Warspite*:	Commissioned 18/04/67	4500 tons dived (PWR 1 Core A)
HMS *Churchill*:	Commissioned 15/07/70	4500 tons dived (PWR 1 Core A)
HMS *Conqueror*:[a]	Commissioned 09/11/71	4500 tons dived (PWR 1 Core A)
HMS *Courageous*:	Commissioned 16/10/71	4500 tons dived (PWR 1 Core A)
HMS *Swiftsure*:	Commissioned 17/04/73	4500 tons dived (PWR 1 Core B)
HMS *Sovereign*:	Commissioned 11/07/74	4500 tons dived (PWR 1 Core B)
HMS *Superb*:	Commissioned 13/11/76	4500 tons dived (PWR 1 Core B)
HMS *Sceptre*:	Commissioned 14/02/78	4500 tons dived (PWR 1 Core B)
HMS *Spartan*:	Commissioned 22/09/79	4500 tons dived (PWR 1 Core B)
HMS *Splendid*:	Commissioned 21/03/81	4500 tons dived (PWR 1 Core B)
HMS *Trafalgar*:	Commissioned 25/05/83	5200 tons dived (PWR 1 Core Z)
HMS *Turbulent*:	Commissioned 28/04/84	5200 tons dived (PWR 1 Core Z)
HMS *Tireless*:	Commissioned 50/10/85	5200 tons dived (PWR 1 Core Z)
HMS *Torbay*:	Commissioned 07/02/87	5200 tons dived (PWR 1 Core Z)
HMS *Trenchant*:	Commissioned 14/01/89	5200 tons dived (PWR 1 Core Z)
HMS *Talent*:	Commissioned 12/05/90	5200 tons dived (PWR 1 Core Z)
HMS *Triumph*:	Commissioned 02/10/91	5200 tons dived (PWR 1 Core Z)

(*continued*)

G. M. Jones, *The Development of Nuclear Propulsion in the Royal Navy, 1946–1975*, Security, Conflict and Cooperation in the Contemporary World,
https://doi.org/10.1007/978-3-031-05129-6

continued

Ship Submersible Nuclear—SSN		
HMS *Astute*:	Commissioned 27/08/10	7400 tons dived (PWR 2 Core H)
HMS *Ambush*:	Commissioned 01/03/13	7400 tons dived (PWR 2 Core H)
HMS *Artful*:	Commissioned 18/03/16	7400 tons dived (PWR 2 Core H)
HMS *Audacious*:	Commissioned 03/04/20	7400 tons dived (PWR 2 Core H)
Ship Submersible Ballistic Nuclear—SSBN		
HMS *Resolution*:	Commissioned 30/10/67	8400 tons dived (PWR 1 Core A)
HMS *Renown*:[a]	Commissioned 15/11/68	8400 tons dived (PWR 1 Core A)
HMS *Repulse*:	Commissioned 28/09/68	8400 tons dived (PWR 1 Core A)
HMS *Revenge*:[a]	Commissioned 04/12/69	8400 tons dived (PWR 1 Core A)
HMS *Vanguard*:	Commissioned 14/08/93	15,900 tons dived (PWR 2 Core G)
HMS *Victorious*:	Commissioned 07/01/95	15,900 tons dived (PWR 2 Core G)
HMS *Vigilant*:	Commissioned 02/11/96	15,900 tons dived (PWR 2 Core G)
HMS *Vengeance*:	Commissioned 21/11/99	15,900 tons dived (PWR 2 Core G)

[a]To fit in with the timescale of the Polaris programme, two Polaris and one Hunter-Killer submarines were built by Cammell Laird at Birkenhead

Appendix B

Initial Dreadnought ME Crew

Samborne, B. F. P., Lt Cdr	Trained in USS *Skipjack*
Squires, R., Lt Cdr	Trained in USS *Skate/Swordfish*
Grove, J., Lt Cdr	Trained in USS *Skate*
Hammersley, P. G., Lt Cdr	Trained in USS *Skipjack*
Hutchinson, C. H., Lt Cdr	
Manson, J., Lt	
King, B. F., Lt	
Cochrane, I., Lt Cdr	
Timmis, K. P. I., Lt Cdr	
Bowyer, W., CPO	Trained in USS *Halibut/Sargo*
Everett, G., CPO	
Faulkner, T., CPO	Trained in USS *Skipjack*
Flavell, D., CPO	
French, C., CPO	Trained in USS *Halibut/Sargo*
Lemmon, E., CPO	Trained in USS *Skipjack/Skate*
Maryon, I., CPO	Trained in USS *Skipjack*
Moorhouse, P., CPO	
Potter, A., CPO	
Rawle, D., CPO	Trained in USS *Skate*
Rudkin, C., CPO	
Treen, P., CPO	
Walls, H., CPO	Trained in USS *Halibut/Sargo*
Welford, CPO	
Young, K., CPO	Trained in USS *Halibut/Sargo*

G. M. Jones, *The Development of Nuclear Propulsion in the Royal Navy, 1946–1975*, Security, Conflict and Cooperation in the Contemporary World,
https://doi.org/10.1007/978-3-031-05129-6

Glossary

ADEB Admiralty Development Establishment Barrow
AEO Admiralty Engineer Overseer
AERE Atomic Energy Research Establishment (Harwell)
AEX Atomic Energy Executive
AFO Admiralty Fleet Order
AGR Advanced Gas Reactor
ANP Advanced Nuclear Plant
APWP Atomic Propulsion Working Party
ARL Admiralty Research Laboratory
ARTE Admiralty Reactor Test Establishment
ASW Anti-Submarine Warfare
AWRE Atomic Weapons Research Establishment (Aldermaston)
BJSM British Joint Services Mission (Washington)
BSRA British Shipbuilders Research Association
CBNS Commander British Navy Staff (Washington)
CORDEP Core Development Programme
CERA Chief Engine Room Artificer
CPO Chief Petty Officer
CSA Chief Scientific Advisor
DCNS Deputy Chief of Naval Staff

G. M. Jones, *The Development of Nuclear Propulsion in the Royal Navy, 1946–1975*, Security, Conflict and Cooperation in the Contemporary World,
https://doi.org/10.1007/978-3-031-05129-6

DEE Department of Electrical Engineering
DEPC Defence Equipment Policy Committee
DGS Director General Ships
DLE Dreadnought Liaison Engineers (US Team)
DMEO Deputy Marine Engineering Officer
DNC Director Naval Construction
DNES Director Naval Education Service
DNP Director Nuclear Propulsion
DPT Dreadnought Project Team
DRPC Defence Research Policy Committee
DSIR Department Scientific & Industrial Research
DS/MP Dounreay Submarine Prototype
E-in-C Engineer in Chief
EOP Emergency Operating Procedure
FBD Fast Battery Drive
FLIP Forward Looking Investigatory Programme
FOS/M Flag Officer Submarines
GUPPY Greater Underwater Propulsive Power
HSD High Separation Diffusion
HTP High Test Peroxide (H_2O_2)
IBR Integral Boiling Reactor
JCAE Joint Committee on Atomic Energy
MAUD Military Application of Uranium Detonation
MCP Main Circulating Pump
MDA Mutual Defence Agreement Act (1958)
MEO Marine Engineering Officer
MoD Ministry of Defence
MoS Ministry of Supply
MPCP Main Propulsion Committee Panel
NATO North Atlantic Treaty Organisation
NPB Naval Projects Board
NPTR Naval Prototype Test Reactor
NSC Naval Safety Committee
NSR Naval Staff Requirement
NST Naval Staff Target
NuDe(S) Nuclear Development (Submarines)
NWSC Nuclear Warships Safety Committee
ORNL Oak Ridge National Laboratory
PWR Pressurised Water Reactor

RANP Rear Admiral Nuclear Propulsion
RCNC Royal Corps of Naval Constructors
RFA Royal Fleet Auxiliary
RNSS Royal Naval Scientific Service
RPV Reactor Pressure Vessel
RR&A Rolls-Royce & Associates
S5W Submarine 5th Generation Westinghouse
SDPC Ship Design Policy Committee
SEO Staff Engineering Officer
SGHW Steam Generating Heavy Water
SHP Shaft Horse Power
SIP Secondary Plant Improvement Programme
SIR Submarine Intermediate Reactor
SOP Standard Operating Procedure
SPSC Ship Propulsion Sub Committee
SSB Ship Submersible Ballistic
SSBN Ship Submersible Ballistic Nuclear
SSG Ship Submersible Guided Missile
SSGN Ship Submersible Guided Missile Nuclear
SSK Ship Submersible Conventional
SSN Ship Submersible Nuclear
STR Submarine Thermal Reactor
SWR Saturated Water Reactor
TA Tube Alloys
TCE Technical Chief Executive
UKAEA United Kingdom Atomic Energy Authority
USAEC United States Atomic Energy Commission
VNEL Vickers Nuclear Engineering Limited
WSAP Westinghouse Special Atomic Programme
Y-ARD Yarrow Admiralty Research Department

Bibliography

Primary Sources

Private Papers:

Berry, Roger, Commander RN Ret'd
Bowyer, W., Ex CERA
Farley-Sutton, Colin, Captain RN Ret'd
Hammersley, Peter, Rear Admiral RN
Hill, Sir Robert, Vice Admiral RN
Jacobsen, John, Captain RN Ret'd

MoD Abbey Wood, Director of Nuclear Propulsion (DNP):

DNP 1, ALO 02617/97 – 000378 Vol I First Lord's (Galbraith) Committee on the application of nuclear power to marine purposes, Caveat: Top Secret Case/ Secret Atomic
DNP 2, NP184/2011 The Dreadnought Project Outline Narrative, Caveat: Secret
DNP 3, CP60801/1956 [New NP185/2011] Machinery of a submarine propelled by nuclear power. Agreement with UK Atomic Energy Authority for the Authority's services in connection with, Caveat: Secret

G. M. Jones, *The Development of Nuclear Propulsion in the Royal Navy, 1946–1975*, Security, Conflict and Cooperation in the Contemporary World,
https://doi.org/10.1007/978-3-031-05129-6

DNP 4, N.8 [New NP186/2011] Naval reactor programme and research and development, Caveat: Secret
DNP 5, N/CP24 Rolls-Royce & Associates Ltd, Constitution, Caveat: Secret
DNP 6, CP60801 UKAEA Collaboration with Admiralty and its contractors in RN nuclear propulsion programme, Caveat: Secret

The National Archives (TNA):

Records of the Ministry of Supply/United Kingdom Atomic Energy Authority

AB 1/313B, Use of atomic energy for power by the Admiralty, 1945–47
AB 4/1014, Report by the MAUD Committee on the use of Uranium as a source of power, 1941
AB 6/81, Liaison with the Admiralty, 1946–48
AB 6/567, Thermal fission reactor: Admiralty reactor shielding, 1948–52
AB 6/618, Liaison with Admiralty, 1949–51
AB 6/626, Enriched Reactor Group, 1949–51
AB 6/747, Boiler design for marine application: collaboration with the Admiralty, 1947–51
AB 6/751, The enriched reactor: correspondence and meetings with Metropolitan Vickers Co. Ltd, 1950–52
AB 6/760, 'N' class submarines: discussions with the Admiralty, 1950
AB 6/1051, Liaison with Admiralty, 1952–56
AB 6/1269, Feasibility study for the Mk I enriched reactor by Metropolitan-Vickers Electrical Co. Ltd, 1950
AB 6/1270, Feasibility study for the Mk I enriched reactor by Metropolitan-Vickers Electrical Co. Ltd, 1951
AB 6/1271, Feasibility study for the Mk I enriched reactor by Metropolitan-Vickers Electrical Co. Ltd, 1952
AB 6/1599, Liaison with Admiralty Research Laboratory concerning the future of Naval Section staff, 1955–59
AB 6/1737, Access Agreements: Admiralty submarine project; Rolls Royce, 1955–57
AB 6/1818, Liaison with the Admiralty, 1956–61
AB 6/1867, Nuclear submarine (Dreadnought) project: organisation, 1956–58
AB 6/1987, US/UK exchange of reports on nuclear marine propulsion, 1957–61
AB 6/2308, Nuclear submarine (Dreadnought) project organisation, 1959–60
AB 6/2492, Admiralty Submarine Programme: organisation and review, 1960–65
AB 8/791, Technical assessment: Dreadnought reactor, 1957–59

AB 8/966/1, Part I – Marine reactor development: Dreadnought programme, 1959–63
AB 8/966/2, Part II – Initial problems of the PWR design and related experimental programme, 1959–63
AB 8/1126, Technical assessment: Dounreay Submarine Prototype Reactor, 1960–61
AB 15/2043, An enriched uranium reactor for submarine propulsion Author(s): J. Diamond J. Smith, 1951
AB 15/2186, Metropolitan-Vickers Electrical Co. Ltd.: Enriched reactor Mk I feasibility report, Section I, 1952
AB 15/5939, Report on the visit of Admiral Rickover and party to the Metallurgy Division AERE, 1956
AB 16/1379, Marine propulsion: naval reactor for Admiralty submarine, 1954–57
AB 16/1380, Marine propulsion: naval reactor for Admiralty submarine, 1957–62
AB 16/1764, Marine propulsion: naval nuclear submarine project; discussions with Admiralty, 1956
AB 16/2032, Nuclear Submarine Committee, 1957
AB 16/2742, United States of America – United Kingdom Collaboration on submarine reactor, 1958–59
AB 16/3162, Nuclear powered submarines; exchange of information with United States of America, 1956–57
AB 16/3287, Nuclear ship propulsion: naval vessels including submarines, 1956–63
AB 64/280, Rolls Royce and Associates (RRA) and Y-ARD work submissions: NUDE(S) II design and development programme for naval reactor plant, 1966–69
AB 89/75, Dreadnought nuclear submarine project: organisation, 1965–68

Records of the Admiralty:

ADM 1/14995, Question of transfer of scientists from centimetre valve team to Tube alloys group, 1943
ADM 1/17080, Request from Sir John Anderson for release of certain staff by Admiralty Scientific Research Establishments for tube alloy work in America, 1944
ADM 1/20414, Functions of operating submarines: investigation and report into possibilities of operating submarines as convoy escorts, or with hunting groups in a fighter role, 1947–49; 1954
ADM 1/22451, Development of machinery for nuclear submarines, 1951–52
ADM 1/23729, Requirements for an HTP operational submarine, 1952–53
ADM 1/25252, Submarine versus submarine attacks: report of trials, 1952–53
ADM 1/26740, Fuel elements for nuclear submarine: Admiralty Board decision, 1957

ADM 1/26756, Operation RUM TUB: analysis, 1958
ADM 1/26843, HM Submarine Dreadnought: staff requirements, 1957–58
ADM 1/26860, Nuclear fuel submarines, 1950–53
ADM 1/27088, Nuclear propulsion: Naval personnel training, 1958–59
ADM 1/27198, Correspondence with Admiral, British Joint Services Mission (ABJSM), Washington, about visit of Vice Admiral Rickover US Navy to United Kingdom, February 1959, 1958–59
ADM 1/27375, Nuclear submarine and POLARIS Project, 1957–58
ADM 1/27750, Submarine machinery future Research and Development, 1960–61
ADM 1/27946, Machinery for nuclear submarine: future research and development, 1961–67
ADM 1/28870, RN College, Greenwich: major works items to provide nuclear propulsion training facilities, 1963–64
ADM 1/29274, HMS/m Dreadnought (SSN 01): Lessons learnt during construction, 1963–64
ADM 116/6204, Atomic reactors: exchange of information with the USA, 1956–57
ADM 116/6411, Nuclear propulsion unit for HMS/m Dreadnought, 1958
ADM 205/102, Admiralty, Office of the First Sea Lord, 1952–54
ADM 205/106, Admiralty, Office of the First Sea Lord, 1955
ADM 205/112, Admiralty, Office of the First Sea Lord, 1955–56
ADM 205/178, Admiralty, Office of the First Sea Lord, 1957–58
ADM 247/79, Note on the possibilities of applying atomic energy to submarine propulsion, 1947
ADM 287/40, Electrolytic oxygen production for breathing on nuclear powered submarines, 1957
ADM 317/204, Advanced Nuclear Plant (ANP) prototype secondary plant: termination report, 1976
ADM 328/29, Rolls Royce/Westinghouse contract, RRA/Electric Boat: Refuelling HMS/M Dreadnought, 1965–75

Records of the Cabinet Office:

CAB 21/4059, Collaboration with the USA on research and development, 1945–54
CAB 21/4852, Marine nuclear propulsion, 1957–61
CAB 126/173, Admiralty interest in nuclear energy for ship propulsion, 1944–46

Records of the Ministry of Defence:

DEFE 4/204, Chiefs of Staff Committee: Minutes of Meetings 40-46, 1966
DEFE 5/47, Chiefs of Staff Committee: Memoranda Nos 291-377, 1953

DEFE 7/2055, Nuclear propulsion for ship and submarine, 1950–58
DEFE 13/523, Minister's Office. American offer of submarine nuclear propulsion unit, 1958
DEFE 19/50, Chief Scientific Adviser. Nuclear Propulsion, 1957–59
DEFE 19/243, Arrangements for re-fuelling HMS Dreadnought, 1963–66
DEFE 69/749, Dreadnought: negotiation of purchase of US propulsion plant, 1957–58
DEFE 72/45, Relations with the USA stemming from the Bi-Lateral Agreement (Cmnd 537 and 733) under which the provision of HMS Dreadnought's machinery and other benefits were supplied to the UK, 1954–68

Records of the Medical Research Council:

FD 1/468, Research workers, 1946–50

Records of the Foreign Office:

FO 800/854, Discussions on atomic energy at meeting of Dominion Prime Ministers in London, May 1946

Records of the Ministry of Power:

POWE 14/1157, United Kingdom Atomic Energy Authority: Admiralty submarine reactor, 1958
POWE 74/83, Rebuilding of Jason at Royal Naval College, Greenwich, 1960–68

Records of the Prime Minister's Office:

PREM 8/1097, Minister of Defence forwarded a note about his talk with Senator McMahon on Atomic Energy development in U.S.A., 1949
PREM 8/1162, Defence research and production programmes 1949–1950: Prime Minister's anxiety that the problem of submarine menace should be considered jointly by Navy and Air Force; service production programmes 1950–1951; disclosure of defence information, 1949–50
PREM 8/1244, Proposed development of a nuclear fuel submarine: Prime Minister's approval for statement by First Lord of the Admiralty in the House of Commons, 1950
PREM 11/2554, UK-US co-operation in the field of atomic energy, 1957–58
PREM 11/2635, Development of nuclear submarine: agreement on Anglo-US co-operation, 1957–59

Records of HM Treasury:

T 225/1022, Development of nuclear submarines, demand for fissile material, 1951–56
T 225/1159, Admiralty: new construction programmes, 1955–60
T 225/1191, Development of a nuclear submarine, including purchase of machinery from USA: fissile material for programme on nuclear propulsion for staff, 1958
T 225/1192, Development of a nuclear submarine, including purchase of machinery from USA: fissile material for programme on nuclear propulsion for staff, 1958
T 225/1193, Development of a nuclear submarine, including purchase of machinery from USA: fissile material for programme on nuclear propulsion for staff, 1959
T 225/2206, Admiralty: Development of nuclear submarine prototype at Dounreay Scotland, 1960–64
T 225/2663, Design, construction and modernisation of the Nuclear Submarine HMS Dreadnought, 1960–66
T 225/2964, Nuclear submarine construction programme, 1965–66
T 225/3324, Nuclear submarine construction programme, 1966–69

Records of the Works Department and Ancient Monuments Board:

WORK 14/2691, Adaptions to King William Buildings for Department of Nuclear Science, 1953–64

Southampton University, Broadlands Archive – Earl Mountbatten of Burma (MB):

MB1/I300, First Sea Lord's Quarterly Newsletters
MB1/I391, Duncan Sandys, PC, MP, Minister of Defence 1957–59
MB1/I397, Sea Lords' meetings 1957–59
MB1/I447, Admiral Arleigh A. Burke US Navy, Chief of Naval Operations 1957–59
MB1/I478, Visit to Canada and the United States of America
MB1/J40, Admiral Arleigh A. Burke US Navy 1959–65
MB1/K208A, Letter to P. Hammersley 31 July 1979
MB1/N104, Admiral Hyman G. Rickover USN 1957–59

London School of Economics, Archive of the Campaign for Nuclear Disarmament:

CND/1/68, Polaris, 1962–68

Parliamentary Command Papers:

Third report from the Select Committee on Estimates; 1946–47 (132)
Admiralty statement of the first lord of the Admiralty. Explanatory of the navy estimates; 1947–48 (Cmd. 7054)
Admiralty statement of the first lord of the Admiralty. Explanatory of the navy estimates; 1948–49 (Cmd. 7337)
Department of Scientific and Industrial Research: Report for the Year 1947–8, With a Review of the Years 1938–48; 1949 (Cmd. 7761)
Admiralty statement of the first lord of the Admiralty. Explanatory of the navy estimates; 1950–51 (Cmd. 8160)
Admiralty statement of the first lord of the Admiralty. Explanatory of the navy estimates; 1951–52 (Cmd. 8476)
Navy dockyard and production accounts 1952–53. Annual accounts of (a) ship building and ship repairing in H. M. dockyards and by contract, and of other dockyard transactions and (b) production at other establishments, for the financial year 1952–53, together with the report of the comptroller and auditor general thereon; 1953–54 (119)
Articles of Agreement governing collaboration between the authorities of the U.S.A. and the U.K. in the matter of Tube Alloys* [*Atom Bomb Research and Development]; 1953–54 (Cmd. 9123)
Admiralty. Explanatory statement on the navy estimates 1956–57 by the first lord of the Admiralty; 1955–56 (Cmd. 9697)
United States No. 2 (1956) Amendment to agreement between the government of the United Kingdom of Great Britain and Northern Ireland and the government of the United States of America for co-operation on the civil uses of atomic energy of June 15, 1955; 1955–56 (Cmd. 9789)
Defence: Outline of Future Policy; 1956–57 (Cmnd. 124)
Treaty Series No. 41 (1958) Agreement between the government of the United Kingdom of Great Britain and Northern Ireland and the government of the United States of America for co-operation on the uses of atomic energy for mutual defence purposes; 1957–58 (Cmnd. 537)
United Kingdom Atomic Energy Authority. Seventh annual report for the period 1st April 1960–31st March 1961; 1960–61 (229)
Navy Estimates 1963–64; 1962–63 (113)

Parliamentary Debates—Hansard:

Lords Sitting of 10th May, 1950, 1950; House of Lords Hansard, George VI year 14; Columns 225–288, Fifth series Volume 167

Lords Sitting of 11th April, 1951, 1950–51; House of Commons Hansard, George VI year 15; Columns 229–290, Fifth series Volume 171

Lords Sitting of 27th March, 1952, 1951–52; House of Lords Hansard, Elizabeth II year 1; Columns 1023–1098, Fifth series Volume 175

Lords Sitting of 27th July, 1954, 1953–54; House of Lords Hansard, Elizabeth II year 3; Columns 121–190, Fifth series Volume 189

Commons Sitting of 8th March, 1956, 1955–56; House of Commons Hansard, Elizabeth II year 5; Columns 2279–2486, Fifth series Volume 549

Commons Sitting of 14th May, 1958, 1957–58; House of Commons Hansard, Elizabeth II year 7; Columns 21–36, Fifth series Volume 588

Commons Sitting of 11th July, 1963, 1962–63; House of Commons Hansard, Elizabeth II year 12; Columns 1385–1606, Fifth series Volume 680

Articles:

Journals General:

Anon, 'Atomic Energy Research Establishment Progress at Harwell', *British Medical Journal,* (31 July, 1948), 263–265

Cockcroft, J. D., 'The possibilities of nuclear energy for heat and power production', *Proceedings of the Institution of Mechanical Engineers* (1947), 206–211

Daniel, R. J., 'The Royal Navy and Nuclear Power', *Transactions of the Institution of Naval Architects,* Vol. 90, No. 4, (Oct., 1948), 273–290

Edwards, Professor. J., 'Joint Panel on Nuclear Submarine Propulsion: Initial Problems of the Submarine Pressurized Water Reactor Design and the Related Experimental Programme', Lecture to the Institute of Marine Engineers, 23 January 1962

Goldberg, Alfred, 'The Atomic Origins of the British Nuclear Deterrent', *International Affairs (Royal Institution of International Affairs 1944-),* Vol. 40, No. 3, (Jul., 1964), 409–429

Gordon, Robert B., 'Miracle Metals Serving in War and Peace', *Power Engineering,* Vol. 104 (10), (Oct., 2000a), 118–128

Jukes, J. A., 'Nuclear Energy: A Survey of Britain's Position', *International Affairs (Royal Institution of International Affairs 1944-),* Vol. 32, No. 3, (Jul., 1956), 273–282

Martin, Laurence W., 'The Market for Strategic Ideas in Britain: The "Sandys Era"', *The American Political Science Review,* Vol. 56, No. 1, (Mar., 1962), 23–41

Menke, J. R., 'Nuclear Fission as a Source of Power', *Econometrica*, Vol. 15, No. 4, (Oct., 1947), 314–334

Owen, Sir Leonard, 'Nuclear Engineering in the United Kingdom – the first ten years', *Journal of British Nuclear Energy,* (Jan., 1963), 23–32

R. G. H., 'The Nuclear Attack Submarine', *The Naval Review,* Vol. LV, No. 1, (Jan., 1967), 22–26

Rogers, Robert B., 'Trends in Soviet Naval Strategy', *Naval War College Review*, Vol. 21, No. 6, (Feb., 1969), 13–29

Smith, C. Jay and Ernest J. King, 'The World War II Soviet Navy Revisited: A 1965 evaluation and a look at the future', *Naval War College Review*, Vol. 18, No. 7, (Mar., 1966), 10–51

Smyth, H. D., 'Fifty years of Atomic Physics', *Proceedings of the American Philosophical Society,* Vol. 90, No. 1, (Jan., 1946), 1–6

Starks, J. F., 'German "U"-Boat Design and Production', *Transactions of the Institution of Naval Architects,* (1948), 291–315

Thomson, George, 'Nuclear Energy and its uses', *International Affairs (Royal Institution of International Affairs 1944-),* Vol. 22, No. 3, (Jul., 1946), 315–325

Thomson, George, 'Anglo-U S. Cooperation on Atomic Energy', American Scientist, Vol. 41, No. 1, (Jan., 1953), 75–80

Weinberg, A. M., 'High pressure water as a heat transfer medium in nuclear power plants', (Atomic Energy Commission Report MonP-93, 10 April 1946) http://web.ornl.gov/info/reports/1946/3445605714956 [Accessed 24 Aug. 2015]

Journal of Naval Engineering (JNE)

Anon, 'Notes from sea', *Journal of Naval Engineering,* Vol. 8, Book 1, (Jan., 1955), 67–80

Anon, 'Notes from sea', *Journal of Naval Engineering,* Vol. 10, Book 2, (Apr., 1957), 231–243

Anon, 'FASMAT Nuclear Submarine Simulator', *Journal of Naval Engineering,* Vol. 19, Book 3, (Jun., 1971), 482

Bailes, R. M., "H.T.P.' For Submarine Propulsion', *Journal of Naval Engineering,* Vol. 7, Book 3, (Jul., 1954), 301–303

Beard, C. S., 'Correspondence: Some Thoughts on Engineer's Office Organization', *Journal of Naval Engineering,* Vol. 11, Book 3, (Dec., 1958), 452–460

Brown, D. K., 'Submarine Pressure Hull Design and Diving Depths Between the Wars', *Journal of Naval Engineering,* Vol. 30, Book 3, (Dec., 1987), 511–519

Clack, N. B. M., 'Practical Training of Nuclear Submarine Power Plant Operators', *Journal of Naval Engineering*, Vol. 19, Book 1, (Jun., 1970), 25–30

Edwards, J., 'Nuclear Courses at the Royal Naval College', *Journal of Naval Engineering*, Vol. 19, Book 1, (Jun., 1970), 15–24

Hairs, E. T. D., 'The Arctic Snort Patrol of H. M. S/M. "Ambush"', *Journal of Naval Engineering*, Vol. 2, Book 4, (Jan., 1949), 391–394

Lees-Spalding, I. J., 'Better Late Than Never', *Journal of Naval Engineering*, Vol. 16, Book 1, (Dec., 1965), 73–78

Manhire, B. D., 'Contamination of Gun Mounting Hydraulic Systems', *Journal of Naval Engineering*, Vol. 9, Book 2, (Apr., 1956), 214–219

Russell, D. H., 'The Cruise of the "Alliance"', *Journal of Naval Engineering*, Vol. 2, Book 3, (Oct., 1948), 288–291

Sampson, W. H., 'A.S.R.1', *Journal of Naval Engineering*, Vol. 9, Book 01, (Jan., 1956), 56–65

Sanderson, A. S. C., 'German Hydrogen Peroxide U-Boat Engine', *Journal of Naval Engineering*, Vol. 1, Book 1, (Apr., 1947), 84–86

Sims, Sir Alfred J., 'Some post-war developments in naval construction', *Journal of Naval Engineering*, Vol. 13, Book 2, (Dec., 1961), 174–205

Taylor, L. A., 'Submarine Engineering Problems and Development 1939–1945', *Journal of Naval Engineering*, Vol. 2, Book 2, (Jul., 1948), 115–135

Trapnell, A. H. L., 'Standardization in the drawing office', *Journal of Naval Engineering*, Vol. 15, Book 1, (Jun., 1954), 148–159

Trewby, G. F. A., 'H.M.S. Sultan: The Marine Propulsion Machinery School', *Journal of Naval Engineering*, Vol. 14, Book 3, (Dec., 1963), 315–333

Wilson, M. A., 'Nuclear Submarine Reactor Plant – Practical Training: Admiralty Reactor Test Establishment, Dounreay', *Journal of Naval Engineering*, Vol. 17, Book 1, (Jun., 1967), 7–15

Journal of the Royal Naval Scientific Service

Adkins, B. M., 'The Atomic Energy Research Establishment, Harwell', *Journal of the Royal Naval Scientific Service*, Vol. 3, (1948), 189–194

Edwards, J., 'Nuclear Power Plants: An Outline of the Experimental Programme Necessary to Support a Nuclear Power Project', *Journal of the Royal Naval Scientific Service*, Vol. 11, (1956), 255–278

Edwards J. and J. E. R. Holmes, 'Neptune: The Admiralty Zero Reactor', *Journal of the Royal Naval Scientific Service*, Vol. 13, (1958), 223–232

Kendrick, A. L., 'The R.N.S.S. and Neptune, Part 1', *Journal of the Royal Naval Scientific Service*, Vol. 14, (1959), 226–236

Secondary Sources

Books:

Anon, *Atomic Challenge: A Symposium* (London, Winchester Publications Limited, 1947)

Anon, *Submarines of The Royal Navy* (Liskeard, Maritime Books, 1982)

Alden, John D., *The Fleet Submarine in the U. S. Navy: A Design and Construction History* (London, Arms and Armour Press, 1979)

Baker, Rowland and L. J. Rydill, 'The Building of the two Dreadnoughts. European Shipbuilding One Hundred Years of Change', *Proceedings of the Third Shipbuilding History Conference at the National Maritime Museum, Greenwich*, 13–15 April 1983 (National Maritime Museum, 1983)

Bartlett, C. J., *The Long Retreat: A Short History of British Defence Policy, 1945–70* (London, Macmillan, 1972)

Baylis, John, *Anglo American defence relations 1939–1980: the special relationship* (London, Macmillan, 1981)

Botti, Timothy J., *The Long Wait: The Forging of the Anglo-American Nuclear Alliance 1945–1958* (New York, Greenwood Press, 1987)

Brimacombe, Peter, *Drake's Drum: A history of the Devonport Naval Base and Dockyard* (Plymouth, Mor Marketing, 1998)

Brown, D. K and George Moore, *Rebuilding the Royal Navy: Warship design since 1945* (London, Chatham Publishing, 2003)

Bud, Robert and Philip Gummett, eds., *Cold War Hot Science: Applied Research in Britain's Defence Laboratories 1945–1990* (Harwood Academic Publishers, 1999)

Burcher, Roy and Louis Rydill, eds., *Concepts in Submarine Design* (Cambridge, Cambridge University Press, 1994)

Churchill, Winston S., *A History of the English-Speaking Peoples Vol. 1: The Birth of Britain* (Electronic edition New York, Rossetta Books, 2013)

Critchley, Mike, *British Warships Since 1945: Part 1* (Liskeard, Maritime Books, 1980)

Critchley, Mike, *British Warships Since 1945: Part 2* (Liskeard, Maritime Books, 1981)

Davies, Gareth, ed., *Glossary of Nuclear Terms* www.burges-salmon.com (2014)

Dilks, David, ed., *Retreat from Power: Studies in Britain's Foreign Policy of the Twentieth Century, Volume Two After 1939* (London, Macmillan, 1981)

Evans, David, *Building the Steam Navy – Dockyards, Technology and the Creation of the Victorian Battle Fleet 1830–1906* (London, Conway Maritime Press, 2004)

Friedman, Norman, *The Postwar Naval Revolution* (London, Conway Maritime Press, 1986)

Gowing, Margaret, *Independence and Deterrence: Britain and Atomic Energy, 1945–52. Volume 1: Policy Making* (London, Macmillan Press, 1974a)

Gowing, Margaret—, *Independence and Deterrence: Britain and Atomic Energy, 1945–52. Volume 2: Policy Execution* (London, Macmillan Press, 1974b)

Grove, Eric J., *Vanguard to Trident: British Naval Policy since World War II* (London, The Bodley Head, 1987)

Hendry, John, *Technological decision making in its organizational context: nuclear power reactor development in Britain* (Cambridge, Management Studies Group, University of Cambridge, 1991)

Hennessy, Peter, *Whitehall* (London, Secker & Warburg, 1989)

Hennessy, Peter—, *Cabinets and The Bomb* (Oxford, Oxford University Press, 2007)

Hennessy, Peter & James Jinks, *The Silent Deep: The Royal Navy Submarine Service since 1945*, (London, Allen Lane, 2015)

Hewlett, Richard G. and Francis Duncan, *Nuclear Navy 1946–1962* (Chicago & London, The University of Chicago Press, 1974)

Hezlet, Vice Admiral, Sir Arthur, *The Submarine and Sea Power* (London, Peter Davies, 1967)

Horne, Alistair, *Macmillan 1957–1986: Volume II of the Official Biography* (London, Macmillan, 1989)

Jordan, John, *Soviet Submarines: 1945 to the present* (London, Arms and Armour Press, 1989)

Kemp, Paul J., *The T-Class Submarine: The Classic British Design* (London, Arms and Armour Press, 1990)

Lambert, Harry, ed., *Rolls-Royce: the Nuclear Power Connection* (Rolls-Royce PLC, 2009)

Macmillan, Harold, *Riding the Storm 1956–1959* (London, Macmillan, 1971)

Pach, Chester J. & Elmo Richardson, *The Presidency of Dwight D. Eisenhower, Revised edition* (Lawrence, University Press of Kansas, 1991)

Pocock, Rowland, *Nuclear Ship Propulsion* (London, Ian Allen, 1970)

Pollard, Sidney and Paul Robertson, *The British shipbuilding industry, 1870–1914* (London, Harvard University Press, 1979)

Polmar Norman, *The Naval Institute Guide to the Soviet Navy Fifth Edition* (Shrewsbury, Airlife Publishing Ltd, 1991)

Polmar, Norman, and Thomas B. Allen, *Rickover* (New York, Simon and Schuster, 1982)

Preston, Anthony, *The World's Great Submarines* (Leicester, Blitz Editions, 1998)

Ranft, Bryan & Geoffrey Till, *The Sea in Soviet Strategy* (Basingstoke, Macmillan, 1989 2nd Edition)

Simpson, John, *The Independent Nuclear State: The United States, Britain and the Military Atom* (London, Macmillan Press, 1983)

Smith, Alex, *Lock up the swings on Sunday* (Durham, The Memoir Club, 1998)

Smyth, Henry D., *Atomic energy for military purposes: the official report on the development of the atomic bomb under the auspices of the United States government, 1940–1945* (Princeton, Princeton University Press, 1945)

Tall, J. J. & Paul Kemp, *HM Submarines in camera: An illustrated history of British submarines, 1901–1996* (Leicester, Blitz Editions, 1998)

Verne, Jules, *Twenty Thousand Leagues Under the Sea* (Butler Brothers, New York and Chicago, 1887) https://www.google.co.uk/books/edition/Twenty_Thousand_Leagues_Under_the_Sea

Walker, Peter M. B., ed., *Chambers Dictionary of Science and Technology* (Edinburgh, Chamber Harrap Publishers Ltd, 2000)

Wettern, Desmond, *The Decline of British Seapower* (London, Jane's Publishing Company Limited, 1982)

Ziegler, Philip, *Mountbatten: The Official Biography* (London, Collins, 1985)

ARTICLES:

JOURNALS GENERAL:

Arnold, Lorna, 'A letter from Oxford: The History of Nuclear History in Britain', *Minerva, A Review of Science, Learning and Policy*, Vol. 38, (2000), 201–219

Badash, Lawrence, Elizabeth Hodes, Adolph Tiddens, 'Nuclear Fission: Reaction to the Discovery in 1939', *Proceedings of the American Philosophical Society*, Vol. 130, No. 2, (Jun., 1986), 196–231

Baylis, John, 'The 1958 Anglo-American Mutual Defence Agreement: The Search for Nuclear Interdependence', *The Journal of Strategic Studies*, Vol. 31, No. 3, (Jun., 2008), 425–466

Bernstein, Barton J., 'The Quest for Security: American Foreign Policy and International Control of Atomic Energy, 1942–1946', *Journal of American History*, Vol. 60, No. 4, (Mar., 1974), 1003–1044

Goodman, Michael, 'With a Little Help from My Friends: The Anglo-American Atomic Intelligence Partnership, 1945–1958', *Diplomacy and Statecraft*, Vol. 18, No. 1, (Mar., 2007), 155–183

Hall, Brian, 'HMS *Explorer/Excalibur*, 1956–62', *The Naval Review*, Vol. 94, No. 2, (May, 2006), 132–142

Hill, Sir Robert, 'Forty years on: The UK Naval Nuclear Propulsion Programme', Address to the Institute of Marine Engineers, 10 October 1995.

Hill, Sir Robert–, 'Admiral Hyman G Rickover USN and the UK Nuclear Submarine Propulsion Programme', *International Journal of Naval History*, Vol. 4, Iss. 2, (Aug., 2005), 1–16 The 77th Thomas Lowe Gray Memorial Lecture, IMechE, 19 April 2005

Hudson, George E., 'Soviet Naval Doctrine and Soviet Politics, 1953–1975', *World Politics,* Vol. 29, No. 1, (Oct., 1976), 90–113

Jones, Mark C., 'Give Credit Where Credit Is Due: The Dutch Role in the Development and Deployment of the Submarine Schnorkel', *The Journal of Military History*, Vol. 69, No. 4, (Oct., 2005), 987–1012

Lautenschlager, Karl, 'The Submarine in Naval Warfare, 1901–2001', *International Security,* Vol. 11, No. 3, (Winter, 1986–1987), 94–140

Llewellyn-Jones, Malcolm, 'Trials with HM Submarine Seraph and British Preparations to Defeat the Type XXI U-Boat, September-October 1944', *The Mariner's Mirror,* Vol. 86, No. 4, (Nov., 2000), 434–451

Minter, William and Elizabeth Schmidt, 'When Sanctions Worked: The Case of Rhodesia reexamined', *African Affairs,* Vol. 87, No. 347, (Apr., 1988), 207–237

Redford, Duncan, 'The 'Hallmarks of a First-Class Navy': The Nuclear Powered Submarine in the Royal Navy 1960–1977', *Contemporary British History,* Vol. 23, No. 2, (Jun., 2009), 181–197

Uhlig Jr., Frank, 'The Submarine, 1776–1918', *Naval War College Review*, Vol. 57, No. 2, (Spring 2007), 146–158

Journal of Naval Engineering (JNE)

Brown, David K., 'Sir Rowland Baker, KB, RCNC: A personal appreciation', *Journal of Naval Engineering,* Vol. 36, Book 2, (Dec., 1996), 327–348

Horlick, Sir Ted, 'Submarine Propulsion in the Royal Navy', *Journal of Naval Engineering*, Vol. 27, Book 1, (Jun., 1982), 1–22 The 54th Thomas Lowe Gray Memorial Lecture, IMechE, 26 January 1982

McHugh, C. S., 'H. M. S. 'Sultan' 25 years of Marine Engineering Training', *Journal of Naval Engineering*, Vol. 26, Book 2, (Jun., 1981), 167–185

Wrobel, P. G., 'U.K. Nuclear Submarines: The development of the overall design', *Journal of Naval Engineering,* Vol. 28, Book 1, (Dec., 1983), 106–119

Oxford Dictionary of National Biography (ODNB)

Dalitz, R. H., Peierls, Sir Rudolph Ernst, '*Oxford Dictionary of National Biography*, Oxford University Press, 2004; online edn, https://doi.org/10.1093/ref:odnb/60076 [Accessed 12 July 2020]

Fox, Robert, 'Gowing, Margaret Mary (1921–1998)', *Oxford Dictionary of National Biography*, Oxford University Press, 2004 http://www.oxforddnb.com/view/article/71257 [Accessed 21 October 2015]

'Fuchs, (Emil Julius) Klaus (1911–1988)', Oxford Dictionary of National Biography, Oxford University Press, 2004 https://doi.org/10.1093/odnb/9780192683120.013.40698 [Accessed 3 February 2018]

Lang, Sir John Gerald (1896–1984), *Oxford Dictionary of National Biography*, Oxford University Press, 2004 http://www.oxforddnb.com/view/10.1093 [Accessed 29 June 2018]

Ludlow, N. Pier., 'Sandys, (Edwin) Duncan, Baron Duncan-Sandys (1908–1987)', *Oxford Dictionary of National Biography*, Oxford University Press, January 03, 2008 http://oxforddnb.com/view/10.1093/ref:odnb/9780198614128.001.0001/odnb-9780198614128-e-39858 [Accessed 17 August 2018]

Unpublished PhD Thesis and Other Works

Crowe, William J., 'The policy roots of the modern Royal Navy, 1945–63' (unpublished PhD thesis, Princeton University, 1965)

Gordon, Robert B., 'Working for Admiral Rickover' (unpublished memoir, Naval Historical Foundation Memoir Program, Washington D.C., 2000b)

Llewellyn-Jones, Malcolm, 'The Royal Navy on the Threshold of Modern Anti-Submarine Warfare, 1944–1949' (unpublished PhD thesis, University of London, 2004)

Online Sources

Anon, HMS *Dreadnought* and the Falkland Islands Deployment, 1977 http://www.theguardian.com/uk/2005/jun/01/argentina.military [Accessed 3 December 2015]

Anon, Prestwick air disaster, December 1954 https://en.wikipedia.org/wiki/1954_Prestwick_air_disaster [Accessed 16 March 2016]

Anon, British atomic bomb project http://www.atomicheritage.org/history/British-atomic-bomb-project [Accessed 4 November 2017]

Anon, Admiral William Crowe USN http://www.theguardian.com/news/2007/oct/24/guardianobituaries.usa [Accessed 8 August 2018]

Anon, Dounreay Decommissioning https://www.gov.uk/government/news/shutdown-of-shore-test-facility-reactor [Accessed 21 February 2019]

Anon, A2W https://www.globalsecurity.org/military/systems/ships/systems/a2w.htm [Accessed 18 September 2020]

Anon, Jason casts a cloud over naval college sale https://www.independent.co.uk/news/uk/home-news/jason-casts-cloud-over-1578790.ht [Accessed 18 February 2021]

BBC Genome Project https://genome.ch.bbc.co.uk

British Newspaper Archive https://www.britishnewspaperarchive.co.uk

British Pathe https://www.britishpathe.com

French Submarine Service https://www.britannica.com/topic/gymnote [Accessed 8 November 2021]

Gosling, F. G., *The Manhattan Project: Making the Atomic Bomb* (United States Department of Energy, Washington D.C., 2010) https://www.energy.gov/Im/doe-history [Accessed 19 July 2020]

Herken, Greg, 'London: The Dog That Didn't Bark', in *The Winning Weapon: The Atomic bomb in the Cold War, 1945–1950,* 43–68, (Princeton University Press, 1981) http://www.jstor.com/stable/j.ctt7ztvzm.8 [Accessed 14 July 2020]

HMS Dreadnought Association https://www.hmsdreadnought.website/ [Accessed 25 January 2019]

Holmquist, Carl O. and Russell S. Greenbaum, 'The Development of Nuclear Propulsion in the Navy', *United States Naval Institute Proceedings*, Vol. 86/9/691 (Sept., 1960), 65–71 https://www.usni.org/magazines/proceedings/1960/september [Accessed 11 July 2020]

Nesvizhevsky, Valery, Jacques Villain, 'The discovery of the neutron and its consequences (1930–1940)', *Comptes rendus de l'Academie des sciences throughout history*, Vol. 18, Issues 9–10, (Nov.-Dec., 2017), 592–600 https://www.directscience.com/science/article [Accessed 10 July 2020]

NQ 1 Steel – Department for Business, Energy & Industrial Strategy, (Steel Public Procurement) https://assets.publishing.service.gov.uk [Accessed 4 December 2020]

Nuclear Submarine Launch dates, https://en.wikipedia.org/wiki/list_of_nuclear_submarines [Accessed 15 January 2021]

The Times Digital Archive https://www.gale.com/intl/c/the-times-digital-archive

Turner, Julian, Deep impact: inside the UK's new Successor-Class Nuclear Submarine (29 July 2013) https://www.naval-technology.com/features/feature-nuclear-submarine-successor-uk-royal-navy [Accessed 2 February 2019]

United Kingdom Energy Consumption 2019 https://www.gov.uk/government/statistics/electricity-chapter-5 [Accessed 24 October 2020]

United States Library of Congress https://www.loc.gov [Accessed 23 August 2019]

United States Newspaper Archive https://www.newspaperarchive.com [Accessed 23 August 2019]

Waller, Derek, The U-Boats that Surrendered. U-1407 (HMS Meteorite) in the Royal Navy – 1945 to 1949 https://uboat.net/articles/97.html [Accessed 27 October 2017]

Wells, H. G., *Anticipations of the Reaction of Mechanical and Scientific Progress upon Human Life and Thought* (Chapman and Hall, London, 1901) https://en.wikipedia.org/wiki/Anticipations [Accessed 9 February 2021]

Index[1]

[1] Note: Page numbers followed by 'n' refer to notes.

G. M. Jones, *The Development of Nuclear Propulsion in the Royal Navy, 1946-1975*, Security, Conflict and Cooperation in the Contemporary World,
https://doi.org/10.1007/978-3-031-05129-6

GPSR Compliance
The European Union's (EU) General Product Safety Regulation (GPSR) is a set of rules that requires consumer products to be safe and our obligations to ensure this.

If you have any concerns about our products, you can contact us on

ProductSafety@springernature.com

In case Publisher is established outside the EU, the EU authorized representative is:

Springer Nature Customer Service Center GmbH
Europaplatz 3
69115 Heidelberg, Germany

www.ingramcontent.com/pod-product-compliance
Ingram Content Group UK Ltd.
Pitfield, Milton Keynes, MK11 3LW, UK
UKHW020834290926
15470UKWH00004B/1344
* 9 7 8 3 0 3 1 0 5 1 2 8 9 *